Property Va

ONE WEEK LOAN

C⟨ ⟩ ⟨approach⟩ in more detail while
in ⟨ ⟩ ⟨Scarrett's⟩ established text examines
th⟨ ⟩ ⟨five⟩ principal approaches: comparative,
in⟨ ⟩ ⟨profits and contractors'⟩ methods. The format of the first edi-
tio⟨ ⟩ ⟨maintained;⟩ each method is introduced, explained and then
de⟨monstrated⟩ with worked examples, followed by commentary where appropriate. The concentration on one method at a time enables students and others to go
directly to the method of interest at the time.

Property Valuation is essential reading for students studying valuation as part of
general practice, estate management or many specialized financial degree courses.
It will be of particular assistance to first degree holders in other disciplines entering the profession and wishing to qualify through a conversion diploma or Masters
course. Planners, building and quantity surveyors and practitioners in other related
areas will find sections of the book useful in understanding the process.

Douglas Scarrett is now retired from his last post as Director of Estate
Management at De Montfort University where he was previously Deputy Head of
the School of Land and Building Studies. He entered the educational field after
extensive professional experience in commercial and government organizations
and as an equity partner in an old established firm.

He was joint founding editor of the *Journal of Property Management* and has
been an external tutor for the College of Estate Management for the past 25 years.

Property Valuation

The five methods

Second edition

Douglas Scarrett

Routledge
Taylor & Francis Group

LONDON AND NEW YORK

First published 1991 by E & FN Spon, an imprint of Chapman & Hall

Second edition published 2008 by Routledge
2 Park Square, Milton Park, Abingdon, Oxon OX14 4RN

Simultaneously published in the USA and Canada by Routledge
270 Madison Avenue, New York, NY 10016, USA

Routledge is an imprint of the Taylor & Francis Group, an informa business

© 1991, 2008 Douglas Scarrett

Typeset in Times by GreenGate Publishing Services, Tonbridge, Kent

Printed and bound in Great Britain by TJ International Ltd, Padstow, Cornwall

British Library Cataloguing in Publication Data
A catalogue record for this book is available from the British Library

Library of Congress Cataloging in Publication Data
Scarrett, Douglas, 1929–

Property valuation: the five methods/Douglas Scarrett. – 2nd ed.

p. cm.

Includes bibliographical references and index.

1. Real property – Valuation. I. Title.

HD1387.S28 2008

333.33'2–dc22

2007047910

ISBN10: 0–415–42325–2 (hbk)
ISBN10: 0–415–42326–0 (pbk)
ISBN10: 0–203–96181–1 (ebk)

ISBN13: 978–0-415–42325–0 (hbk)
ISBN13: 978–0-415–42326–7 (pbk)
ISBN13: 978–0-203–96181–0 (ebk)

To Beryl,
for her steadfast support

Contents

Figures

Tables

Preface

The five methods of valuation are well established approaches to the valuation process and together provide the basis for valuations for a wide range of purposes. They are methods that were developed in the United Kingdom and are now used in most parts of the developed world.

Many of the processes can be performed by preset computer programs, but still rely on the knowledge and expertise of the valuer in providing sound information for the constituent parts.

The ability to support, explain and defend the opinions and judgements contributed to the valuation is vital. A valuation for sale, purchase, compulsory purchase, business accounts or rating may well be challenged and needs to be supported by clear and conclusive evidence, often in the face of a paucity of relevant information on recent transactions. This approach was well received in the first edition. For these reasons, I have continued in this new edition to explore each method in turn and to describe in some detail the process and the underlying reasons.

The defence to any suggestion of carelessness or incompetence consists of the notes made by the valuer and supported by the evidence on which the inputs were based. The need to maintain comprehensive notes on the thought processes and the evidence on which they were based as an integral background to the published valuation cannot be overemphasized.

Some attempt has been made to incorporate recent shifts in treatment, and it is hoped that the new text will be of assistance not only to valuation students but also to practitioners and members of allied professions such as building surveying and planning (and graduates from other disciplines joining the profession and discovering the subject for the first time).

I acknowledge the support and advice of Matthew Smith during the early stages of preparation of this second edition. I have benefited directly and indirectly in discussions with colleagues to whom I give my grateful thanks. I wish to thank my sons, and in particular Paul for his contribution to the task of assembling the chapters and undertaking the presentation of examples, figures and tables.

Gretcha De Burca, Manager, RICS Scotland Library and Information Services has been unfailingly helpful and resourceful.

The support and encouragement of Kate McDevitt, Eleanor Rivers and latterly Katy Low at publishers Taylor & Francis is gratefully acknowledged.

Abbreviations

AIM	Alternative Investment Market
ASP	alternatively secured pension
AVSC	Asset Valuation Standards Committee
BCIS	Building Cost Information Service
CMP	Code of Measuring Practice
DCF	discounted cashflow
DRC	depreciated replacement cost
ERV	estimated rental value
FRI	full repairing and insuring
FRV	full rental value
FTSE 100	Financial Times Stock Exchange top 100 companies
FTSE 250	Financial Times Stock Exchange top 250 companies
FV	future value
GDC	gross development cost
GDV	gross development value
GEA	gross external area
GIA	gross internal area
HIPs	Home Information Packs
ICT	information and communication technology
IPD	Investment Property Databank
IR	internal repairing
IRFY	interest risk-free yield
IRR	internal rate of return
ISA	individual savings account
ITZA	in terms of Zone A
IVSC	International Assets Valuations Standards Committee
LSE	London Stock Exchange
MLR	minimum lending rate
NIA	net internal area
NPV	net present value
OFEX	Off Exchange
p.a.	per annum
PACT	Professional Arbitration on Court Terms

PEPs	personal equity plans
psm	per square metre
PV	present value
REITs	real estate investment trusts
RICS	Royal Institution of Chartered Surveyors
RPI	Retail Price Index
RV	rateable value
SDLT	stamp duty land tax
SF	sinking fund
SIPPs	self-invested personal pensions
SSSI	Site of Special Scientific Interest
TEGOVA	European Group of Valuers Associations
TESSAs	tax-exempt special savings accounts
USP	unsecured pension
VAT	Valued Added Tax
VOA	Valuation Office Agency
YP	years' purchase

Acknowledgements

My thanks to the publishers E & FN Spon for permission to include an edited extract from one of the price schedules from *Spon's Architects' and Builders' Price Book* included in Appendix A.

Thanks also to The Royal Institution of Chartered Surveyors which owns the copyright of the definition of market value on page 43, reproduced with permission.

1   Setting the scene

Any but the most rudimentary exchange system requires measurement, assessment and permanent records, so these skills developed in relation to land at an early stage in history.

The English legal system developed from the feudal order imposed following the Norman Conquest. Thus the Domesday Book, completed in 1086, recorded the property holdings, values, rights and obligations of the King, his tenants-in-chief, their sub-tenants and peasants, while a court roll contained similar information for each manorial estate. The information collected was invaluable in economic and political terms. Such records are necessary for the collection of taxes and to facilitate action deemed to be necessary by the governing body.

Formal assessments of land value became necessary at least as early as 1695, when an Act of Parliament provided for a rental multiplier to be used as the basis of compensation. The occasion was the demolition of weirs to enable improvements to be carried out to natural rivers to improve their suitability as navigation channels for inland transportation.

There are now only two legal estates in land: freehold and leasehold, as provided in the Law of Property Act 1925. Freehold – the fee simple absolute in possession – while nominally held of the Crown, in practice confers absolute ownership, enabling holders to do whatever they please with the land subject to the rights of others and the laws in force at the time. Planning laws provide the main constraints on change.

Leasehold is a term of years absolute, conferred by the landlord on the tenant. It is more often referred to as a lease or tenancy, which sets out the duration, rent and other terms agreed between the parties and detailing their rights and responsibilities. Unless a lease for any term of seven years or more is granted formally (that is, by deed), the effect will be to create an equitable interest only.

Eventually, the property statutes of the 1920s brought about reform, but land tenure has inherent complications that require detailed provision. Tithes were abolished in 1936. The payments were converted into annuities for 60 years, the process therefore being completed in 1996. Provision was made for the enlargement of certain long leases of residential property into freehold interests despite the wishes of the freeholder, on the payment of compensation which, except in the case of leases with very short terms unexpired, amounted to little more than nominal sums.

Records of transactions and encumbrances have been transformed under the Land Registry, making use of sophisticated computerized systems. Any sale must now be registered, with the result that all real estate will have a registered title in due course.

POST-WAR DEVELOPMENT

At the end of the Second World War, the property market was in a dormant state. The only construction activity allowed during the war period and for some time afterwards was to provide essential buildings. Few repairs had been done and many buildings, both domestic and commercial, especially those in areas judged to be vulnerable to attack, had been boarded up. The building industry recovered only slowly; materials were in short supply. New housing could be built only under licence unless it consisted of replacing buildings destroyed or extensively damaged by enemy action.

The immediate problem after 1945 was the need to decide what to do with the major areas affected by bomb damage. Some city centres, including Plymouth, Bristol and Liverpool, had suffered heavy and sustained bombing. A few centres were redeveloped quickly, facilitated by the availability of substantial areas of space following wholesale destruction. The layout was largely in the style of the earlier pattern and was accessed directly from busy streets. Pedestrianized centres were introduced later, often with an upper floor element, which rarely proved popular. The covered centre, with lifts, escalators, integral car parking and public transport terminus, and a comprehensive management and maintenance system came much later.

At the same time, provision of housing was regarded as a public sector priority, with the local authorities being the only major providers. To this end, the phrase 'working classes' used in the Housing Act 1936 was dropped from the 1949 Act of the same name. Even including the contribution of the private sector, where the houses were almost entirely for owner occupation, central government found difficulty in approaching its overall target of 500 000 units a year.

For some time after the war, the property market remained to a large extent localized, low key and fairly predictable. Much commercial and industrial property was owned, not as an investment but as a business asset by the owner. The property investment scene has changed out of all recognition since those times.

Meanwhile, the Town and Country Planning Act of 1947 made a bold but ultimately unsuccessful attempt to deal with the increases in land value brought about by a provision for the expropriation of private development rights in return for a claim on a 'once for all' notional global sum of compensation of £300 million.

A rigid code of land use was introduced in 1948, giving planning a largely restrictive role in the development process while exerting a major influence on land values, ensuring widespread use of the planning appeal procedure.

THE GROWTH OF PROPERTY INVESTMENT

The private sector quickly recognized the potential for multiplying the value of suitable development land assisted by low cost, non-equity capital readily supplied by the banks and others. It was often possible to borrow the whole of the development costs in the belief that their security was sound, based on the anticipated value of the completed scheme.

The local authority structure was unsuited, and its financial resources unequal, to the task of commercial development, although many tried to invigorate their town centres by providing mainly small schemes, often associated with multi-storey car parking. Other authorities preferred not to be involved directly, disposing of areas of land carefully pieced together by the use of compulsory powers to commercial groups anxious to exploit the development potential.

Some councils retained an element of interest by granting a building lease and receiving an income from the scheme. The overriding advantage of this was the ability to control redevelopment of the site in the longer term. Bristol, Coventry and many other local authorities structured their disposals in that way. Coventry has recently upgraded its shopping centre. Meanwhile, Broadmead (Bristol) is about to have expenditure of some £500 million on an ambitious project to re-establish its importance as a regional shopping centre in the face of substantial competition from Cribbs Causeway and elsewhere.

It is difficult to understand now how much the activities of a few nationally known property developers such as Levy, Hyams, Clore and Cotton dominated not only the financial pages of the broadsheets but also the news pages of the popular press. They were variously acclaimed as financial wizards or denigrated as parasites depending on the observer's point of view. Undoubtedly they had imagination and flair but were gradually overtaken: as schemes became larger, property companies adopted more formalized structures and became more accountable to their shareholders.

It was not until the mid 1960s that insurance companies began to look for a share in the equity of schemes for which they had provided much of the venture capital to the increasingly large national property investment companies. Various forms of sharing were devised, a popular one being the provision of funds at rates below market level in exchange for a share in the equity or even for the right to purchase the whole scheme based on an agreed formula related to the eventual rent roll, once it had been completed and let successfully. In such cases, the developer acted as a catalyst, using the company's expertise in the development process in return for a capital profit – a form of entrepreneurial project manager.

The first motorway was opened in 1959 and others followed together with outer and inner ring roads for many towns and cities – not all completed, even now.

Market activity continued and intensified to a point where, in the early 1970s, there was a spectacular market crash. Some depositors and shareholders, anticipating the seriousness of the situation, hastened the crisis by transferring cash to the established banks (at the time described as 'a fit of collective prudence'). The crash affected not only the property companies but extended to the whole of the secondary banking system to such an extent that the Bank of England intervened with a financial lifeboat to mitigate the worst effects of the substantial losses suffered. Liability was shared with the major clearing banks but the details were shrouded in secrecy. It was some time before confidence returned to the property market.

Eventually it did so, with the players pursuing increasingly ambitious and imaginative town centre schemes, large-scale office developments and out-of-town business parks. The network of motorways has opened up the possibility of viable regional shopping centres on a vast scale, accessible by millions of visitors living within little more than an hour's journey on the motorway.

The discounted sell-off of council housing started slowly but soon gathered pace, especially for the better designed units in good locations. The rump not readily sold found its way into housing stock held and managed by housing associations as most local authorities were anxious to divest themselves of ownership and responsibility. Some of the original purchasers eventually made use of the equity provided by discounted prices and a rising market to move up the property ladder.

A more recent trend has been the purchase of houses and flats to let, facilitated by the removal of major restrictions on rent levels and security of tenure beyond that agreed contractually between the parties. As a result, there is a thriving market in the purchase of residential units to be let as investments. Some investors have borrowed more than is prudent, leaving themselves exposed to financial risk should the property be vacant for any length of time. The cost of borrowing will not be treated as a business expense unless the loan is on an interest-only basis. The additional demand created mainly at the lacklustre lower end of the market may have contributed to rapidly rising prices putting ownership beyond the reach of some who would have expected to purchase their own house.

Housing associations have expanded their operations and also offered some units on a shared ownership basis, where the occupier purchases a share in the property and pays a rent for the remainder. An element of subsidy is included as the accommodation is intended for key workers and for families unable to fund their own provision.

Meanwhile, some local authorities have used their planning powers to restrict development except in specified areas and then to require developers to include a percentage of affordable units in the scheme. It is unlikely that the effect of these latest moves will be more than marginal.

At the same time the overall market has grown. It is inevitable that each successful new development creates displacement elsewhere, so that earlier and more traditional developments of, say, shops or offices may register vacancies. They may remain occupied but the overall quality of the occupiers may be affected. There have also been shifts in fashion, so that a centrally situated warehouse, for example, that was converted into offices in the 1960s but is now deficient in many ways, provides the opportunity for a development of up-mar ket city centre flats.

THE ROLE OF THE VALUER

The valuer is a pivotal figure in the smooth operation of the property market. Any proposal for development (including putting together a viable site from land in a variety of ownerships and legal interests), application for planning permission, negotiation of a loan, buying and selling of an interest, arranging a letting, or settling a rent review to the best advantage is likely to involve one or more valuers, often interacting with other professionals. The ongoing management activity is no less important, ensuring that the investment performs to its full potential and any opportunities to maximize the investment are recognized and reported on.

The valuer is available to advise organizations that own or lease premises from which they operate their business activities. The range of services includes advice on managing existing assets, rental levels at review and on expiry and recommendations about acquisitions and disposals so as to shape the holding to match corporate aspirations. The activity may be carried out by a firm of valuers and surveyors retained to look after all the company's property interests or by various firms engaged to advise and negotiate on a particular matter as it arises. The larger companies tend to have a skilled in-house team, calling on outside firms only where particular expertise or another perspective on the problem is required.

THE FUTURE

The introduction in 2007 of real estate investment trusts (REITs), described elsewhere, should provide opportunities for smaller investors to enter the property sector without the drawbacks that exist in direct ownership.

So, although the market is buoyant and confident, that is not to say there are no problems. Large companies are increasingly locating routine office work and call centres overseas, thereby achieving significantly savings. Home working has not yet taken off, but there are many benefits to both parties from such activity and increasing congestion on roads may speed up the changeover. Congestion charges, discussion about road pricing and other restrictions may affect particular sections of a city or area and spark a reaction. Measures to reduce the effect of climate change may well result in significant changes to current patterns of behaviour. Buying over the Internet is gaining in popularity and having an effect on retailers in the markets where such purchasing is strong, such as books and electronic items.

Some multiple retailers have moved from town centres to business or retail parks to reduce overheads and acquire more display space while at the same time developing an online presence. Departmental stores, especially those on several floors, look vulnerable except where they are in any of the more successful modern retail developments. The established retailers such as Marks & Spencer, Bhs, Debenhams and House of Fraser are under intense pressure from niche fashion chains, well-financed supermarkets and cost-conscious retailers such as Primark and Matalan, which have exploited the low cost of imports to maximize turnover rather than achieve high margins to generate good trading results. As a consequence, future demand for premises for business use may show marked shifts of emphasis, affecting rental and capital values.

All these activities are informed to a greater or lesser extent by the valuation exercise. It is therefore essential that the valuer understands not only the property investment market but also the general investment market and the way in which returns interact. Valuers need to be well equipped to offer high quality service and advice to their clients. To be capable of such a level of service, valuers need to understand the foundations of the valuation methods employed.

This text sets out to explore the structure and rationale of each of the five methods of valuation with a view to facilitating comprehension. To that extent, each section is self-contained, although on occasion further consideration of a topic may be found elsewhere. For example, depreciation and obsolescence are considered specifically in relation to the contractor's approach but also feature in a broader consideration in the chapter on determinants of value.

2 The overall investment market

The valuer has an expert knowledge and awareness of the property market, which has its own peculiarities, especially in the direct investment section, entry to which requires access to substantial funds and a long-term commitment. There are of course opportunities to invest in property companies, but their shares have often traded at a discount to assets. The introduction of real estate investment trusts (REITs) may go some way to opening the market to individual investors, particularly to pension funds.

Some knowledge of the wider investment market will show the main depositories for investors' savings. An appreciation of their attraction and the returns available should serve to extend the valuer's appreciation of the competition faced by the real estate investment market. A summary of some of the main financial products available follows.

CONSUMPTION AND SAVINGS

Every individual needs a certain level of income to ensure a given standard of living. Once this level has been achieved there is a choice – increase consumption (effectively increasing one's standard of living), or put aside some or all of the surplus income to provide for the uncertainties of the future.

Savings accumulated in this way may be put in a box and buried in a field to be retrieved when required later, or allowed to accumulate, earning interest, to be available to purchase an asset at some future date. The type of investment chosen should be one able to fulfil the investor's wishes. The purpose may be one of many, such as to accumulate savings for an, as yet, undefined objective, to save for a specific anticipated expense such as school fees, to provide funds for a major purchase, or to supplement one's income in retirement.

Unless surpluses are invested in some way, their purchasing power will tend to decrease broadly and at least in line with the current level of inflation.

Where a market exists that is both active and well publicized (a perfect market), it produces a pattern of prices that is informative to observers of the market and persuasive to would-be participants in the market. In order for there to be a market, there must be both buyers and sellers. The fact that there are shades of opinion is what constitutes a market. If everyone in the market held the same views, there would be minimal activity. As it is, the prospective investor is able to observe the market and discover the price level and range on any particular trading day and also the number and size of bargains struck at that level. Prices can be monitored over a period, performance compared across the market.

One such market is that in shares. Large companies have millions of shares and thousands – perhaps tens of thousands – of shareholders. When shareholders decide to sell, they may decide to sell a part or all of their holding in a company. Motivation is varied; sellers may need the liquidity offered by a sale or may believe that the share price is higher than is justified by the performance of the company and that a profit should be taken before the share price falls.

The fact that buyers exist shows that not everyone reacts in the same way. While the seller may expect the price to fall, the buyer is presumably optimistic of future performance based on a purchase at the current level.

In the United Kingdom (UK), the main organization for dealing in shares is the London Stock Exchange (LSE). There are also subsidiary markets, the main ones being the Alternative Investment Market (AIM) and the Off Exchange (OFEX).

A capital sum invested in government securities, shares in companies or other forms of savings in the private sector will attract a financial acknowledgement for its use. In the case of government stock, the capital sum will be secure, and as a result payment for its use will tend to be low. In many government-backed initiatives designed to promote longer term savings, such as individual savings accounts (ISAs), the income from cash savings and other specified forms of investment will be tax-free, although there are restrictions on the amount invested in any one year. Withdrawals are permitted but cannot be replaced at a later date. The maximum annual amounts qualifying for such treatment are fairly modest.

In the private sector, there is a more direct relationship to the market. The rate of interest on investments in building societies and other savings agencies will be influenced by the Bank of England's minimum lending rate as well as the demand for loans. Building societies compete with each other on mortgage loan rates so will not pay a rate of interest greater than that necessary to ensure a sufficient inflow of funds for the level of business that they wish to achieve.

The dividend on ordinary shares will relate to the trading profits of the company and its need to hold back some of the surplus to bolster reserves and fund investment to keep its operations up to date and competitive. Where no profit has been made, no dividend will be available unless the company has sufficient reserves and decides to draw on them.

Where savings are directed towards the provision of a pension fund to commence at some future date, annual earnings on the investment will be consolidated into the capital fund to swell the amount eventually available to fund an annuity or otherwise provide a retirement income. One of the attractions of saving in this way is the tax relief on payments into the fund. However, having received favourable tax treatment, there are restrictions on how the fund is eventually used to provide a pension.

Those already well provided for and not needing to secure an income could consider the purchase of a work of art or a piece of antique furniture. These are not investments for the ordinary saver with more mundane aspirations and limited resources.

Even for those people with the ability and desire to save there must be an expectation of clear future benefits. The recent failures of endowment policies to mature at the levels forecast when they were taken out has affected confidence in this type of saving. Lower interest rates and the decision to tax pension fund income have contributed to shortfalls in many pension schemes and the demise of the pension based on final salary for many in the private sector.

The alternative available is to retain assets in a realizable form. Where high interest rates are offered it may be that the interest on short-term money deposits is greater than the return on conventional investments. This strategy is a holding one and is unlikely to be viable except in the short term. Inflation will diminish the purchasing power of the liquid assets, eroding the real return.

The minimum lending rate

The minimum lending rate is important because it is one of the benchmarks against which returns are judged.

Traditionally, monetary policy in the UK was controlled by government. The Bank of England's base rate was used to reflect political considerations and to control economic growth.

The Bank of England was nationalized in 1946. In 1997, the government relinquished its overt control over determination of what is now termed the minimum lending rate. The rate is fixed at a level recommended by the bank's Monetary Policy Committee. The committee consists of nine members, each with one vote. Five members are drawn from the Bank of England, including the governor and two deputy governors. The remaining four members are appointed by the Chancellor of the Exchequer to serve for a three-year term, which can be renewed. The nine members are charged with setting rates for the cost of borrowing and the return on savings to control inflation at an annual rate fixed by the government. The target rate of inflation in 2007 was 2 per cent as measured by the consumer price index, but may be varied. A deviation of 1 per cent or more either way triggers a requirement for the chairman of the bank to explain the circumstances to the Chancellor. The committee meets every month to determine the changes, if any, and the amount. The reform has been welcomed because it makes the process transparent through the published minutes. It also ensures, in theory at least, that action will be taken as soon as it is necessary rather than be influenced and possibly delayed by political considerations.

The market significance of the minimum lending rate is that it will act as an indicator of the return expected by investors prepared to look further afield and accept an element of risk. Some investors will be prepared to shoulder what they perceive as a low level of risk in exchange for a better return: the level of risk will be judged by individual investors. They may also decide to invest in those types of investment where the reward comes partly from the annual return and partly from an increase in the value of the asset in which the investment is made. For example, a company having a good product with increasing turnovers will be in a position to declare a good dividend while the success of the company will be reflected in the higher value of shares on the stock market, enabling the investor to sell and take a profit on the capital originally invested.

THE COMPONENTS OF A SOUND INVESTMENT

Every investment has strengths and weaknesses. The investor will decide on a case-by-case basis the trade-off between return and risk, and will wish to have regard to some fundamental points in determining which types of investment to select. The main attributes sought in an investment may be summarized as follow:

- easily and speedily bought and sold, without restrictions on access and with low dealing and transfer costs
- a positive income, i.e. one that at least offsets the effect of inflation
- homogeneous and divisible
- well and fully definable and documented
- one for which there is a demand
- minimal management
- not politically sensitive
- prospect of increase in capital value.

The absence of any of these qualities will not necessarily exclude the investment under consideration from inclusion in the group from which the selection will be made, but should guide the investor in his or her choice. The alternative is to look for other forms of investment that accord more closely with the main require-ments of the investor.

More specifically, the investor will consider each of the following points in deciding which investment has the greatest overall attraction.

Yield

The total yield is made up of a risk-free return and a risk premium. The risk-free return reflects the yield on government stock that is regarded as the universal ref-erence point of safe investment. Yields are also influenced by the perceived risk of the particular investment. The risk premium attempts to measure the market response to the exposure to a particular level of risk. The yield will take account of annual interest payments or dividends, as well as any capital growth achieved on the eventual sale or maturation of the investment.

Return

The incidence of receipt of interest payments is important. Payment made quarterly is worth more than payment made annually, where the nominal rate is the same. The actual rate is slightly higher where payment is made quarterly, as the sums are available for reinvestment at an earlier date.

In the event, most investors arrange their affairs by allocating some of their savings to accounts repayable on demand without penalty and some to longer term investments where there is no prospect of needing the money quickly; the sacrifice of that facility would be expected to improve the rate of return. A further sum may be placed in relatively risky investments, but such decisions should be determined in answer to the question: 'Can I afford to lose this amount of money?'

Risk and security

The ordinary meaning of risk may be defined as the extent to which the outcome diverges from the expectation. The outcome may be greater than expected, which will be welcomed by the investor, but may also be less than expected, which will be a source of concern. There is also a relative risk, where the income has a lower purchasing power than anticipated.

As well as being secure, the income needs to be sufficient to justify the investor's decision to forgo immediate access to and use of capital in favour of the income anticipated. There must be some reasonable prospect that the income will not reduce significantly in real terms over the foreseeable future.

Liquidity and divisibility

Some investments are more easily and readily tradable than others. To take two extremes, stocks and shares can be bought and sold in a day or two and the account settled in two weeks; interests in real estate may take some months to find a buyer and some little time beyond that to complete the legal transfer, while until contracts are exchanged the transaction may be aborted by either party. Similarly, the difficulty of offering investments in appropriate lots is much easier with shares than with property, where there is the problem of indivisibility. In the latter case, disposal of an investment to meet a cash requirement may result in a surplus needing to be reinvested, with the attendant risk of not finding a suitable investment.

The investor will expect to be able to realize any investment within a minimum time scale. Unexpected events happen in people's lives and it is important to be able to realize assets quickly where the crisis is of a financial nature. It follows that any penalty for early redemption needs to be considered very carefully.

Management

All investments require management, and if this is entrusted to professional managers, it will carry a fee. Some investments may need only periodical reviews to consider whether they continue to offer the best return available in that investment class for the risks accepted. Physical investments, such as property, will need a permanent management oversight and result in a higher fee. For example, tenancy changes, rent reviews, advice on insurance cover, condition reports and ongoing investment direction will all incur charges that, although difficult to quantify in advance, need to be provided for.

Dealing costs

When investing (or liquidating) capital, an investor wishes to see the whole amount paid or repaid. Any fees for arranging the deal or for closing the account will erode the actual rate of return. Where there are significant areas in which advice is needed, fees paid for expert advice should be accepted as necessary; they may save the investor from expensive mistakes.

THE SOURCE OF INVESTMENT FUNDS

Funds for investment in commerce and for the public expenditure programme of the state are provided by the public through savings and taxation.

Few private investors have access to sufficient funds to enable them to acquire substantial amounts of stocks, shares or other investments directly, but indirectly their individually modest contributions to pension funds and insurance policies provide vast sums.

As personal wealth and influence have been replaced by institutional funds, an increasingly sophisticated and knowledgeable approach to the art of investment has arisen. Institutional investors tend to manage their funds in an active way, being prepared to switch from one holding to another when market indicators demand it.

They are often criticized for taking a short-term view in influencing a course of action to secure a short-term benefit that is not necessarily in the best long-term interests of the company in which the investment is held. The fund manager bases decisions on market research into performance and forecasts prepared by analysts. Institutions act conservatively with regard to risk taking. They avoid the more volatile stock and attempt to build and maintain a balanced portfolio – a range of investments, forming a complementary whole, spreads the exposure to risk and enables them to meet their known future liabilities. Paradoxically, fund managers do take some risks, as evidenced by their investment in Internet shares, which have so far proved highly unpredictable. They have also shown an interest in equity and hedge funds. Many of the larger funds stipulate the balance of shares to be held in different segments of the market; some favour tracking the companies listed in the Financial Times Stock Exchange top 100 UK companies (FTSE 100).

Private equity and hedge funds offer significantly improved potential returns, although they are illiquid and exhibit substantial risk levels.

More recently, international investment has taken on greater significance. Undoubtedly there are opportunities for investment although, as has been seen recently, the legal and regulatory systems of other countries are not always sufficiently robust or mature in protecting the investor. Overseas investment also runs the additional risks of double taxation and exchange rate variations. The potency of some of the world's political systems suggests that such investments should be selected with great care.

Meanwhile, goods will increasingly be made where they can be supplied on the keenest terms, and labour-intensive activities will be carried out in developing economies where labour costs are cheap. The cost of transfer would at first sight appear to limit such changes. But in the global economy, supplies of oil and its derivatives, clothing and even fresh fruit and vegetables are often transported half way around the world without major competitive detriment. As an example, the carriage cost of a single bottle of wine from Australia is less than 1p, hardly significant in the overall price structure. The increasing emphasis on global warming may influence some of these decisions in the future.

Any widespread introduction of congestion charges and road pricing, as seems likely, might well re-order the appeal of some types of property investment and some locations. Perhaps the greatest impact will come from the twin pressures of looming energy shortages and the drive to reduce carbon emissions. Both are likely to involve fundamental changes. The effect on property is unclear at this time but is likely to be far reaching.

THE RANGE OF INVESTMENT OPPORTUNITIES

The investor has a wide spread of investment choice, each with its own characteristics. The main differences are between investments in public stock or commerce, fixed and variable income-producing investments and direct and indirect participation.

FIXED INTEREST INVESTMENTS

Fixed income investments include government stock, company loan stocks and preference shares.

Government stock

Fixed interest

Stock is issued to finance part of the public expenditure. It is referred to as 'gilt edged', reflecting the underlying security afforded by state underwriting. It is inconceivable that the government would default on interest payments or on agreed redemption dates, giving the stock a security status against which all other forms of guarantee tend to be measured.

Government stock is usually issued at a par of £100 nominal value, although the initial price may be above or below the par value; any variation will affect the coupon yield. The effective yield on trading after issue will be calculated by relating the rate of return to the price paid. Most issues have a date that refers to the year when the stock will be redeemed at its par value; a stock with a low interest yield will rise in value as it approaches its redemption date.

An investor wishing to buy or sell stock after it has been issued will not deal with the government but through the stock exchange. The market fixes the price according to the yield attached to the stock and its perception of current yields and trends.

Government stocks are categorized as follows:

- Short dated ('shorts')   redeemable within 5 years
- Medium-dated (mediums)   redeemable after 5–15 years
- Long-dated (longs)   redeemable after 15 years
- Undated (one-way option stock)   (undated or irredeemable).

The undated or option category comprises stock redeemable only at the option of the borrower (the government). Some option stock issues have no date for redemption, while others have a date such as '2015 or after'. Where such stocks are coupled with a low rate of interest, it is unlikely that the government will exercise its option to redeem, and trading in the market will ignore the possibility of redemption in arriving at a market value. A variable rate can be obtained by investing in index-linked stock (described below).

Variable interest – index-linked stock

Index-linked stock is also issued by the government. The sum invested receives an interest payment varying in line with the retail price index (RPI). It has a fixed life at the end of which the nominal value is repaid in full together with the RPI increases. Both interest and capital are index linked.

Traditionally, government has raised funds through the issue of stock with or without a redemption date and at a fixed rate of interest as described earlier. In addition, in recent years it has sought to encourage savings without itself receiving the funds or providing payment for its use. It has done this through a variety of products that are mainly distinguished by some form of favourable tax treatment.

The following products are included under the government section because they benefit from favourable tax treatment.

Individual savings accounts (ISAs)

Individual savings accounts comprise savings that are free of both income and capital gains tax. Limited, fairly modest funds may be invested annually. The funds can be transferred to other providers to optimize return, but once withdrawn cannot be replaced.

There are two types – maxi and mini ISAs. Maxi ISAs permit investment of a maximum annual amount of £7000, which may be held in cash or shares (but with a maximum cash component of £3000). Mini ISAs are limited to an annual investment of £3000, which may be held in cash deposits (or £4000 if invested in stocks and shares).

Personal equity plans (PEPs) and tax-exempt special savings accounts (TESSAs) fulfilled this purpose in the 1990s but both were replaced by ISAs after 5 October 1999.

Some providers offer investments in identified sectors of the market, including overseas holdings. The investor should bear in mind that such acquisitions may involve additional risks and that it would be prudent to take advice before venturing into such commitments.

Self-invested personal pensions (SIPPs)

Funds require careful and skilful management to gain the optimum advantage from the concessions available. There needs also to be a recognition that the purpose of giving tax relief is to provide funds for future pension needs and the use of the assets within the fund are subject to the rules.

Alternatively secured pensions (ASPs)

Unless someone with a private pension purchases an annuity by the age of 75, the fund will be designated as an alternatively secured pension. The advantage is that, unlike an annuity, which terminates on the death of the beneficiary, the remaining funds can be transferred by will. However, such funds will be subject to inheritance tax.

COMMERCIAL INVESTMENTS

Fixed interest securities

Apart from government stock, few investments attract a fixed rate of interest; where they do so, the term is likely to be quite limited and the yield modest. These terms tend to reflect the general inadvisability of guaranteeing a fixed level of payment from income that is not guaranteed and may fluctuate appreciably from one year to the next.

The issues tend to be small and specific and consequently trading is less active.

Loan stock

Companies issue loan stock as one form of share capital, secured on particular assets of the company and rewarded with a fixed rate of interest. Investors are not shareholders and have no voting or other rights in the company.

Debenture stock

Stock is usually secured on specific assets of the company with a further floating charge over the remaining assets of the company. The advantage of debenture stock is its prior ranking over ordinary stock for payment if the company is wound up.

Preference shares

Preference shares usually carry a fixed rate of dividend, payment of which has first call on the company's profits. In the event of liquidation, the preference shareholders rank above the ordinary shareholders for repayment, but after creditors. Preference shares receive more favourable tax treatment than loan stock. Different issues of preference shares may be ranked or treated on an equal basis. Most preference shares are cumulative, which means that a dividend passed in one year is carried forward until it is paid. A non-cumulative preference share loses the payment for any year where the dividend is passed.

Convertible debentures or loan stock attract fixed interest payments but carry rights to convert to ordinary shares at the holder's option, subject to the terms of the issue.

Variable interest investments

The vast majority of investments offer variable rates, leaving the investor with the risk that if the company does not prosper, the shares may not return a reasonable dividend or may not pay any dividend at all in one or more years. This type of investment includes ordinary shares and unit and investment trusts.

Ordinary shares

Companies raise their capital by issuing shares on which a dividend is paid, subject to the profitability of trading (preference shares have fixed returns, as described above). The holders of ordinary shares in companies are the risk takers. When the company is prosperous, the share dividend is likely to increase. But should the company run into difficulties, the shareholders might receive a much smaller dividend or no dividend at all. In the event of the company collapsing, shareholders will be paid only after preference shareholders and creditors have been paid. They may find the value of their investment much reduced or even worthless. Dividends are payable net of tax.

Shares in the top echelon of quoted companies are sought for their dependability and long-term growth, such as those companies included in the FTSE 100. Many funds track the shares of companies and balance their holdings on the performance, involving a good deal of movement at times. There is a similar listing, the FTSE 250, which lists the next best companies. It is suggested that the latter base, composed as it is of sizeable and often well known but generally smaller companies, offers greater diversification and more potential for growth, not least because of the frequent moves in and out of the index.

Unit trusts

For the investor without the financial resources to buy and hold a range of shares over a wide spectrum and/or without the time and expertise to devote to their management, unit trusts offer an alternative. Each trust consists of a form of managed fund safeguarded by a trustee, invariably a bank or an insurance company, which holds the assets, controls the issue of units, maintains a register of holders and oversees the general management. The investor provides funds, which are applied to the purchase of stocks and shares. The share can be encashed at any time although the amount is not guaranteed but based on the value of the fund at that time. The trust will have declared aims and may specialize in certain sectors of the market or certain features. The fund may concentrate on growth or income or pay the same attention to both. The trust has no capital structure and ultimately all the funds including income belong to the unit holders.

Investment trusts

Despite their title, investment trusts are limited liability companies having share capital and the right to issue prior charge capital and to raise new money by rights issues.

Inland Revenue approval of their operation is needed if they are to enjoy the tax advantages usually associated with such trusts. Most of the company's income must come from securities and be distributed; only a limited amount may be retained each year. There is also a limitation on how large a proportion of the funds may be invested in a particular holding. Capital profits must not be distributed. Excessive dealing could endanger the status of the trust.

The trusts cater for private investors who contribute funds to invest in the stock market. They have a fixed number of shares, the value of which varies according to supply and demand.

It is claimed that they incur relatively low management and operating costs, but the real criterion is performance. It is important to select the trust with care, looking at its investment philosophy, management and past record, although it is essential to remember that past performance offers no guarantee for the future achievements of the trust. Trusts have the ability to borrow for further investment in the market. Any shares must be acquired through a stockbroker.

Stakeholder individual savings accounts

Subsequent to the introduction of ISAs, stakeholder ISAs were introduced. They cover a range of financial products that are required to meet certain criteria to ensure that they are uncomplicated and good value. They are offered by savings organisations but benefit from tax concessions granted by the government. There is of course no guarantee of performance.

Self-invested personal pensions (SIPPs)

Annual contributions to a self-invested personal pension scheme enable the investor to control how the payments are invested. The scheme provides for more control and greater flexibility than was previously available in traditional pension arrangements. The maximum permissible annual contribution on which tax relief may be claimed is based on earnings subject to a cap, currently £215 000. There are otherwise no limits on the amount of income allocated to the fund. There are limits on the total capital value of the fund, and if these are exceeded, a punitive penalty of tax for the excess is incurred.

All contributions are made net of basic tax; higher rate taxpayers may claim relief on this part of their contribution also, but in arrears through the tax return. Income from any assets in the scheme is untaxed while any growth is free from capital gains.

A scheme administrator is required, who has the responsibility of ensuring that funds are applied within the regulations governing this type of pension provision.

The member may choose what to invest in from a wide range of assets, including stocks and shares, certain classes of unit trusts, investment trusts, unitized insurance funds and commercial property.

Capital appreciation funds

The principal and usually only return from equity and hedge funds is capital appreciation.

Private equity funds

Institutions, pension funds, other corporate investors and wealthy individuals may all be attracted by the concept of private equity investment. It is essentially a medium- to long-term investment with significant areas of uncertainty. It relies largely on substantial capital appreciation on disposal of an asset. In the normal way, there are no annual payments of interest and there is a higher level of risk of loss of funds depending on the type of investment pursued.

Any private equity investment has limited liquidity; it cannot be traded on the stock market. Typically, an equity fund manager will collect capital commitments from a range of potential investors so as to be in a position to act quickly once an opportunity is found. Any default is likely to be subject to punitive charges. Additional capital is likely to be raised by first offering further investments to existing investors; a well-managed fund with a good track record may be able to operate solely by involving existing investors. Any commitment will be for a maximum amount with the fund manager calling on the whole or part of the amount indicated. Individual amounts are not likely to be less than £500 000 and are more likely to be in the £3–10 million range.

Some funds specialize in providing venture capital to immature companies that have been selected as having promising futures but where there is a substantial incidence of loss. Mezzanine capital provision is less risky as it supports companies that have proven track records but still find traditional fund raising difficult and expensive.

Property, especially property development, is particularly appropriate for this type of funding. A development site with planning permission is more tangible than many other types of investment and is usually capable of providing a probable realization date. Nevertheless, substantial risks remain until the development is complete and the units fully let or sold.

There is no right to require a sale or a return of the investment in advance of completion of a particular project. Quite often, the appropriate time for realization will be apparent.

Distribution of surpluses is available to limited partners only when the particular investment is disposed of. The end payment will be subject to any fees and charges incurred, together with accrued interest.

Hedge funds

Hedge funds are privately owned investment funds that are strongly established in the United States of America (USA) but also operate in Europe, particularly in London. Because of the nature of the market and the restrictions on entry, there is little regulation of funds carrying on their activities either in or from the UK. Moody's operates a subscription rating service that measures past performance and interprets the fund's investment strategy.

Individual funds are often operated as a limited partnership where the partners and managers are also investors. Subject to any limitations imposed by the contract relating to the particular fund, there are no restrictions as to what may be invested in.

The policy of accepting high levels of volatility in pursuit of substantial growth combined with the vast amounts staked involves significant additional risk. There are many investment strategies with varying risk profiles. The funds are aggressive though any risk taking is likely to be combined with actively offsetting the risk. Some funds use derivatives almost exclusively, entering into contracts for differences in preference to purchasing the asset directly, enabling the fund to exert a greater influence on the situation.

A variety of approaches is used, including the following:

- Long/short equity: the fund may buy stock identified by its team as likely to increase in value. But it may also borrow, for a fee, stock that it expects to go down, sell it to third parties and then buy it back at the lower price when it falls in value. Banks sometimes not only lend their stock but transfer their voting rights as well, giving the funds additional leverage. It is significant that such activities may erode a company's share price.
- Trading strategies: taking positions in currencies, markets and commodities.
- Arbitrage: taking advantage of anomalies in the price of assets. The market is not only aware of this tactic but modern research tools and communications make it much easier for any skews to be put back on track at an early date.

- Activist: the acquisition of a stake in a company providing the opportunity for improvement, amalgamation, restructuring or takeover.

Use of the vehicle has demonstrated its ability to make very large profits in a short time although events outside its control may involve large losses. As an example, a US fund was wound up after suffering a loss of some $6 billion when natural gas prices fell against firm expectations of further rises.

The funds receive positive support from some quarters precisely because of their record in improving cashflow and tackling inefficiencies in the operation of companies. On the other hand, there is a view that their activities undermine well-run companies to the detriment of their shareholders.

Hedge funds charge both a management fee and a performance fee, calculated as a percentage of the fund's profit. Early criticism of performance fees led to a more sophisticated approach to reward; in some cases, a performance fee will not be charged until its annualized performance exceeds an agreed level; the fees may be further constrained by a provision to the effect that the manager will not receive performance fees unless the value of the fund exceeds the previous highest net asset value.

The worldwide fall in interest rates has tempted traditional investors such as pension funds trustees and insurance companies to place some of their capital holdings in hedge and equity funds. The sharp fall in interest rates over recent years has affected their ability to garner sufficient income from traditional safe sources to meet their liabilities. Pension funds in particular have been adversely affected by the substantial fall in interest rates, the increase in life expectancy and the unexpected action of government in discontinuing favourable tax treatment of premiums. It is evident that any move in this direction should be well considered and limited in terms of the overall assets and responsibilities of the pension fund.

The attractiveness of these newer types of funds is the opportunity to make large gains against a background of low interest rates and modest yields from more conventional investments.

There is increased pressure to look at whether the tax advantages enjoyed by equity and hedge funds are fair and confer any wider benefits in the community, and an intention to do so was proposed in the October 2007 comprehensive spending review. There is also a concern that many of the companies acquired by the funds and then disposed of have maximized the total profit by failing to make sufficient provision for the liabilities of existing pension schemes.

Venture capital trusts

The venture capital trust scheme was introduced by government in 1995. Venture capital trusts are listed on the London Stock Exchange and are designed to encourage individuals to subscribe to shares where the funds are used to invest in a range of small high-risk companies. The shares and securities of these companies must not be listed on a recognized stock exchange (but may be listed on AIM or OFEX).

When a company becomes quoted on the London Stock Exchange it can continue to qualify for venture trust purposes for up to five years.

The purpose was to facilitate financial loans to small firms that have often in the past experienced difficulty in expanding by the absence of finance on reasonable terms. At the same time, the individual investor benefits from various tax concessions and from the reduced risk because the investment is spread over the range of businesses on which loans have been made.

For a trust to be approved, its income must come wholly or substantially from shares or securities, at least 70 per cent of which must be classified as qualifying holdings; not more than 15 per cent may be invested in any one company, and the trust must not retain more than 15 per cent of its income. The trusts are designed to support a number of companies, enabling the risk to the individual investor to be spread.

Investments are restricted to companies engaged in a qualifying trade, having gross assets not exceeding £7 million prior to any investment by the trust, and not exceeding £8 million immediately after the investment. When a substantial part of the trade is made up of excluded activities, the trust will not qualify. Excluded activities include those dealing in land, financial activities, leasing or letting assets on hire, royalties or licence fees, managing hotels, guest houses, residential care, providing services to another company and farming, market gardening and forestry.

The maximum permitted investment is £200 000, and shares must be held for 5 years to qualify for full tax relief at the rate of 30 per cent on dividends, but only when they are new shares.

Investors will also be entitled to disposal relief on capital gains, where tax will not be due on the sale of shares provided that the trust is approved and the shares were acquired within the maximum yearly limits. Conversely, any loss sustained cannot be claimed or offset against other income.

Investors may subscribe to shares or acquire them through the stock exchange (but some of the benefits are available only on the purchase of new shares).

There are various other provisions to ensure that the scheme is used only for the purpose for which it was designed; as with all schemes where favourable taxation rates and benefits are an important aspect of the decision to invest, the attractions are in the gift of government and may be varied at any time.

REITs offer a new opportunity to invest in property and are described in the next chapter.

Indirect investment

Much indirect investment takes place through contracts of life assurance (entered into to finance home purchase and pensions while providing life cover). The policy may be designed for a specific purpose (e.g. to pay off a mortgage at the end of the mortgage term or provide a pension from a certain age) together with an additional amount payable as a lump sum at the end of the contract. The outcomes are not guaranteed and depend on the performance of the funds in which the premiums are placed. The endowment market has not performed well in recent times and this route is now far less popular than it was.

The investment market is now global. Not only is there strong competition from within the UK for high quality investments, but there is also active interest from overseas investors. There is some support for the weight of money theory that suggests that given the volume of funds seeking a return, there will be a shortage of the highest quality investments, which may result in purchase at a price having little current justification on the basis of the return available.

Other investment media

Apart from property, which is dealt with in the next chapter, another avenue of investment is offered by valuable objects such as pictures, jewellery, antique furniture and works of art generally, not attracting income but expected to show capital appreciation. The increase in capital value needs to be sufficient to offset the lack of income and the cost of keeping the items, such as space, management, maintenance and insurance cover. Antiques or paintings are popular holdings but there is always a risk of an item going out of fashion or deteriorating with age. The purchase of vintage cars is particularly attractive since no capital gains tax is incurred on sale (unless the investor is engaged in dealing).

3 The property market

INTRODUCTION

Property ownership is an important investment medium that has shown substantial income and capital increases over the years. But, unlike stocks and shares, not all property is held for investment purposes. Some is held primarily for the operational purposes of a company, when any increase in value will be incidental. Where such property is retained by the company, its value is restricted by its use for the current purpose. This situation is touched on in later chapters, particularly those describing the contractor's approach to valuation.

This chapter is concerned with the investment potential afforded by property that has its own characteristics and distinctiveness.

In the modern commercial world, there are many calls for the valuation of property holdings. Properties may be occupied by an individual or a company, and the occupiers may be owners or tenants. If the latter, a periodic sum will be paid by way of rent for the use of the premises. There may be several interests in one property – thus a piece of land may be purchased and leased to another on a long-term lease (typically, in this context, 90 or 125 years) with an express undertaking on the part of the lessee to build, say, a shop as specified in the lease. The person building the shop may then occupy it for the purpose of trade or grant a lease for a term of years to a retailer. The retailer may decide to sublet the upper part, fitted out as offices, to a company wishing to operate from that locality. After some time, the retailer may decide to retire and transfer the lease and the goodwill of the business to a third party. Alternatively, the shop could be sublet (subject to obtaining any permission required), in which case the tenant would remain responsible to the landlord for the payment of rent.

A factory owner may purchase a piece of land and erect a factory for the use of the operating company primarily to facilitate its activities. Such arrangements are sometimes left on an informal basis, although this is inadvisable.

It will be seen from these few examples that an interest in property can take many forms. In each case, the need for a valuation of the particular interest held may be required and a valuer will be engaged to carry out the task.

In advising on value, the usual approach is to compare the property to be valued with transactions involving similar properties. The valuer therefore needs a detailed knowledge of the property market or that part of the market under consideration. Some valuers operate in the local area in which they practise. Some, particularly in the provinces, deal with most types of property, while others specialize in one segment of the market, such as warehouses. Sector specialists tend to operate over the whole country, concentrating on, say, prime shops, factories, office blocks or out-of-town developments. They may have a presence in London only, but are more likely to offer a service from the head office supplemented by a number of offices located in the main regional cities. Almost without exception, the leading firms have amalgamated, been taken over or formed alliances with other companies worldwide in an acknowledgement of the global nature of the current property investment scene.

PROPERTY AS AN INVESTMENT

Earlier, the attributes of a perfect investment were considered. Property investment has its own peculiarities, the effects of which are now considered from that perspective.

Yield

The yield from property consists initially of rent paid by a tenant for use of the asset. The landlord will endeavour to pass responsibility for all repairs, maintenance and insurance to the tenant. Not all property is suitable to be let in this way. A tenant may decline to accept the responsibility for all outgoings on an older building. Where the building is nearing the end of its useful life, the landlord may prefer to maintain the same level of rent at the expense of a high standard of maintenance that would perhaps be inappropriate to a building likely to be redeveloped in the foreseeable future.

Where the landlord retains responsibility for some outgoings, the yield is calculated on an estimate of the net rent.

The yield is made up of a risk-free return and a risk premium. The risk-free return has traditionally been related to the return on government stock. The risk premium then involves a calculation on the part of the investor as to the additional return required to accept the level of risk perceived to be inherent in the investment under consideration.

The calculation of risk is cushioned to some extent by the terms of modern leases, in which the downside risk is often protected by an 'upwards only' rent review clause and by the opportunity to renegotiate the rent at each review point and at the end of the lease.

The quality of the investment is all-important in this respect. With modern buildings in sought-after locations, the existing tenant will have to accept that the location is in demand and there will be evidence of recent rents to support any call for an increase. At the end of the lease, should the tenant decide to leave the premises, another tenant is likely to be forthcoming. With a property in a situation where there are vacant premises or insubstantial tenants there is the prospect of no increases on review and the possibility of default by tenants. There may be difficulty in finding new tenants, even at lower rents. Without an opportunity for some form of refurbishment or redevelopment, the future of the investment would look decidedly unattractive.

In recent years there have been attempts to remove or curtail the 'upward only' rent review provision, which is increasingly seen as one-sided and heavily weighted in favour of the landlord. One of the effects of removal would be to hasten the redevelopment of sites, possibly for other uses within a wider regeneration of the surrounding area. It is difficult to defend the upward only review clause that has been a backstop for many years, almost guaranteeing the minimum return from an investment, even where the rental value had fallen below the current level. It could be argued that rents retained at realistic, rather than legalistic, levels would result in fewer defaults by tenants of such properties.

The latest recommendations for a voluntary code are considered later in this chapter.

A further change where an investor needs careful advice follows the loosening of the protection offered to tenants by the Landlord and Tenant Act 1954 as amended. Recent changes have in effect allowed landlords and tenants to agree to enter into a contract outside the provisions of the Act where it suits them to do so, rather than, as previously, having to apply to the courts for permission. The form of the lease will determine the situation between the parties, but it should be understood that any letting outside the Act has implications for both parties. The tenant will not be entitled to compensation if the landlord decides not to renew a lease, and neither could any application be made to the court to fix the terms of a new tenancy. Any wish on the part of the tenant to dispose of the business and recoup goodwill by assigning the lease would not give the assignee any rights to claim a new tenancy following the expiry of the current one. Any prospect of lettings outside the Act would tend to be looked on with some concern by investors and could be reflected in capital values.

Reference was made above to the primary return on the investment depending on the amount of rent received. The overall yield is enhanced when an investment is sold to show an increase over the price paid initially. This feature has had a strong influence on the popularity of investment in property assets over the years, given the sustained growth in property values. It should be noted that any profit would be subject to the provisions relating to capital gains tax.

Ease of dealing

The transfer of property is time-consuming, expensive and, pending exchange of contracts, uncertain. Abortive negotiations are likely to result in unrecoverable costs for both parties.

The sales process consists of a marketing campaign leading to an acceptable offer, after which the legal and financial aspects have to be arranged. As a result, investment in property is described as being relatively illiquid.

There may be considerable delay in completing a sale even when the investment is attractive to the market and terms have been agreed.

Transfer costs include professional fees and stamp duty, the latter currently at rates of up to 4 per cent on the purchase price.

It is generally accepted that property investment requires long-term ownership to maximize benefits, although this concept should be balanced with the needs of portfolio management.

Growth and inflation

The effect of growth and inflation is important. Investment in property has tended to offer steady income growth. When this is anticipated, the initial yield is likely to reflect the prospect. An increase in annual return would tend to have a knock-on effect in the shape of an increased capital value.

The terms of a modern lease ideally provide for regular upward-only rent reviews, full repairing and insuring leases and rent paid quarterly in advance, leaving the landlord with only the costs of management to deduct from the income. The fewer liabilities for outgoings remaining with the landlord, the greater is the certainty of the net income.

Heterogeneity

Each property is unique if only because it stands on a different site. Properties built to similar plans and with similar external appearances may nevertheless have significant differences in layout, use or location. For example, two adjoining and similar houses may have substantially differing interiors, while two structurally identical properties on adjoining sites, one with permission limited to use for retail business and the other only for office use would be likely to have quite different values. An office block in the city centre would be likely to have a different value from a similar building on a business park or on the outskirts of a town.

The uniqueness adds to the care required in the valuation process.

Indivisibility

The problem of indivisibility has long haunted the decision to invest directly in property. When the owner wishes to raise a sum of money for another purpose – say, equivalent to 25 per cent of the capital value of the property – the options are limited. Not only will the sale of the property take some time and incur substantial expenses, but it will also leave the owner with the remaining 75 per cent of the value to be reinvested. A bank loan or mortgage, especially if required for a fairly short term, would be a more realistic approach. It would also retain any tax benefits or allowances accruing from ownership of the property and avoid any assessment for capital gains tax.

Adequate information and a wide market

The property world is not transparent, but limited information about any transaction is now available from the Land Registry on payment of a fee. There are also commercial sources providing details of properties for sale together with further information and analysis of the results achieved.

In addition to the sale price, a valuer needs to have information on the physical attributes, planning constraints and legal status of a property to assist in a proper interpretation of the transaction. A valuer operating in a particular market will have access to formal and informal sources of intelligence and may be able to cross-check to some extent. Any information not capable of verification should be treated with caution.

Demand

Economists point to the inelasticity of supply of property brought about by the extended time scale needed to respond to a perceived increase in demand. It may be possible for individual owners of existing properties to take advantage of the unfulfilled demand in the short term. In the longer term, the market is likely to respond, but provision is uncoordinated and fragmented and if that leads to an eventual oversupply, the investor may be competing in a depressed market to the extent that the return falls short of that anticipated when embarking on the project. The planning control mechanism is not primarily concerned with matching supply and demand.

Costs of management

In comparison with other types of investment, property involves a high management cost. Once acquired, the property needs to be managed competently to ensure that rents are paid on time, covenants are complied with and the property is maintained in good condition. Management is a specialized activity usually entrusted to a firm of surveyors. The firm will also advise on the level of insurance cover and negotiate rents on review and at the end of the lease. The overreaching aspects of management extend to responding to any external occurrences likely to impact on the property, such as proposals for nearby development, road improvements, protection of rights of light and so on. Over the years, there may be opportunities to consider refurbishing or replacing the building, either for the same use or for a more profitable one, or to acquire adjoining premises or sites.

Government intervention

Property is owned within a statutory framework. An investor will be anxious to ensure that there is no likelihood of onerous planning restrictions on the property or changes in the locality that will have an adverse effect on the investment.

There will also be concern about any attempt to restrict the freedom of either party to negotiate freely. Leases of commercial premises are subject to some statutory provisions governing renewal of tenancies, but their balanced nature has ensured a strong market in commercial property investments. In general, a tenant may renew a lease for a further term unless the landlord can show one or more reasons why the lease should not be renewed. Where the landlord establishes the right to possession, there are provisions for compensation. Any improvements authorized by the landlord and carried out by the tenant during the tenancy may also qualify for compensation. There are rules for service of notices, and responses thereto must be complied with. Recent changes relating the market to modern conditions and requirements have resulted in some loosening of the statutory provisions. These are set out in Chapter 5.

NEW VOLUNTARY CODE FOR LEASE TERMS

The majority of leases contain an 'upward only' rent review clause, the effect of which is to ensure that on review, the rent will either rise to reflect current rental values or, if the rental value has not increased or has even fallen below the rent reserved, the rent will remain the same. It is argued that when the rental value has fallen, the tenant should, in fairness, be entitled to a reduction in rent. From the landlord's point of view, the prevailing practice is a valuable one, although difficult to defend on rational grounds. Two previous attempts to introduce greater equality and mutual trust between owners and occupiers in common lease terms have proved unsuccessful. Against this background, the government has hinted that unless progress is made it may seek to change the position by legislation.

A third code has now been launched by the Royal Institution of Chartered Surveyors in the expectation that landlords will accept the proposals in preference to the prospect of legislation to achieve the same ends. The main recommendations cover:

- **Rent reviews:** the majority of leases contain provisions for the rent on review during the course of a lease to be upward only. The code provides that any review should be up or down, subject to the revised rent being not less than the amount agreed between the parties at the commencement of the lease.

- **Assignments and sublettings:** limitations on the provisions relating to assignments and an easing of restrictions on subletting.

- **Break clauses:** the ability to enforce a break clause is often hedged about with requirements for strict compliance with the lease terms, in the absence of which the landlord may refuse consent or at the least cause serious delay. The code requires a more reasonable and consensual approach to dealing with requests to break.

- **Covenant to repair:** whether a full repairing lease is appropriate in any particular case. It may be appropriate to grant leases on modern property on a full repairing lease or subject to a service charge where the landlord takes responsibility for all charges, including fees, which are then allocated in full to the various occupiers. It is less defensible to let an old and dilapidated property on such terms when the tenant is unaware of the full implications of the covenant. Even where a tenant retains professional advisers, there may not be a full realization of the extent to which the repairing obligations involve major expenditure on the structure.

- **Service charges:** the code urges implementation of the Service Charge Code 2002, which emphasizes the need for restrictions on management costs and fees.

- **Consents:** leases provide for the landlord's consent to a wide range of proposed changes, including some that have no adverse effect on the landlord. There is a proposal that consents could be in a standard form. The provision that any request should be dealt with in a reasonable timescale would facilitate progress and possibly reduce the costs involved. It is also made clear that consent is or should be unnecessary in some cases.

- **Insurance:** for good property management reasons, property owners often stipulate that the building insurance shall be taken out with a named company. In such cases, the landlord may qualify for commission on the premium and there is a requirement to disclose the amount of any payment.

RESIDENTIAL PROPERTY INVESTMENT

Residential property is no longer subject to rent restriction or excessive security of tenure. The effect on the sector has been dramatic: there is now an active market in residential letting that supplements the overall supply of housing and is useful to prospective occupiers who cannot afford or have no wish to purchase or will not be in the locality long enough to make purchase a worthwhile route.

The 'buy to let' market appeals both to established investors and to those wishing to gain a foothold in the direct property market. Mortgages are available where the terms require interest payments only, thus reducing the entry level for investors with limited resources wishing to fund repayments from the income derived from the property.

The Inland Revenue is prepared to accept mortgage interest payments as a legitimate expense, but will not allow any part of a repayment mortgage to be reclaimed.

There is some risk when the margins are tight, especially where the property may not be let and therefore not be income-producing. The cost of maintenance is often underestimated, reducing the annual net income. It becomes very easy to fall into arrears with the repayments, especially if the property is available for letting but does not find a tenant immediately.

REAL ESTATE INVESTMENT TRUSTS (REITs)

New rules effective from 1 January 2007 allow property owning companies to manage their tax affairs in a much more tax efficient way. The provisions are similar to those already in place in other parts of the world, particularly France, Australia and the United States of America, where the uniqueness of investment in property is recognized.

Twelve existing companies adopted the scheme at its introduction and more have followed. The new structure gives substantial benefits to investors, which should in time be reflected in share prices.

The new scheme should result in greater freedom to make decisions, particularly about disposal, for sound financial and portfolio management reasons without having to consider the implications of tax. It has long been a feature of property companies that their shares trade at a discount to asset value, a situation that should be eased by the new structure. One immediate effect is that portfolio balance will be more readily achieved.

Subject to provisions designed with a view to preventing the use of the scheme for tax avoidance purposes in close companies, corporation tax will not now be levied on UK profits or on the rental income of companies taking part in the scheme. Further, there is the flexibility to dispose of properties without liability for capital gains. To qualify, a company must have the bulk of its assets – more than 75 per cent – in UK property, a maximum gearing of 1:25, no shareholder owning more than 10 per cent of the company and distribution of at least 90 per cent of its UK profits in share dividends. The entry fee to the scheme is a one-off payment of 2 per cent of the value of the company's UK portfolio.

The provisions have been welcomed by property companies, some having already taken advantage of the new scheme.

Other investment opportunities in real estate

Investors not able or not wishing to invest directly in property may find a suitable property company to purchase shares in, and those companies embracing the REIT provisions set out above should perform better than other property companies.

Funds and trusts (as described in the preceding chapter) provide further opportunities for indirect investment in property, but attention is drawn to the need for substantial capital resources and the higher level of risks involved. Such funds are not suitable if the investor is looking for income returns or certainty in the timing of capital repayments.

4   Valuation mathematics

INTRODUCTION

When the valuer has gathered all the relevant information regarding the interest to be valued, the income flows are quantified to arrive at a capital value. This task is greatly facilitated by the use of the formulae explained below.

Given the option of a sum of money now or at some future date, the rational choice would be to accept it now. Not only is that route clothed with certainty, but there is also then a range of opportunities to spend it on some consumer item or to invest it to gain interest. But if the option is between a certain sum now and a larger sum at some point in the future, the decision becomes one related to the size of the reward for the delay in terms of the interest offered. For example, the offer of £100 now or £105 in one year's time would not be especially interesting to someone who could invest to show a return at 5 per cent (although acceptance now would give the recipient certainty and control of the amount). If, instead of £105, the offer was £103, then the decision would point towards accepting the sum now, whereas an offer of £107 would probably be considered sufficient to defer acceptance in favour of the additional sum later. The exchange rate is the price demanded by an investor in order to forgo current consumption in favour of investing, sometimes referred to as the marginal rate of time preference.

Some results may show slight discrepancies due to rounding, but the amounts involved are infinitesimal and insufficient to affect the integrity of the outcome. A hand-held calculator will give the result to ten significant digits, while published tables are computer-generated and exceed this level (although they are often truncated to keep the printed results manageable). If the output does not exceed seven figures, the degree of accuracy of the last digit is 0.50. The user may be assured that a more than adequate level of accuracy will be achieved whether by calculator or computer.

The need for a representative set of tables was first acknowledged by Richard Parry, a practitioner and partner in the firm of Parry, Blake and Parry, and, later, the first principal of the College of Estate Management. In 1913 he published a set of tables entitled *Parry's Valuation Tables*. The most recent and enlarged edition, the 12th, was published in 2002 under the title *Parry's Valuation and Investment Tables*.

In these days of programmable hand-held electronic financial calculators and personal computers with vast memories there should be no need for a printed set, inevitably incomplete because only a selection of interest rates can be represented. But it is probably still true to say that most valuers possess a copy. Other tables are also published, see Further reading.

The ability to calculate any required result from the relevant formula enables any yield for any period to be obtained.

THE FORMULAE

There are six main formulae, which are shown as reciprocal pairs in the first part of Table 4.1, in which there is the assumption of yearly interest rests. Other formulae for payments in arrears or in advance for periods of less than a year and for aspects of growth are also included.

In each case, the unit is one. The applicable rate of interest ('i') is shown as a decimal (for example 5 per cent becomes 0.05, 11.5 per cent becomes 0.115). The period ('n') between the receipt of one rent payment and the next is in years.

Table 4.1 Valuation formulae

Name and function	Compounding Formula	Name	Discounting Formula
Amount of £1	$(1 + i)^n$	Present value of £1	$(1 + i)^{-n}$
Annuity	$i + \dfrac{i}{(1 + i)^n - 1}$	Years' purchase	$\dfrac{1 - (1/(1 + i)^n)}{i}$
Amount of £1 per annum	$\dfrac{(1 + i)^n - 1}{i}$	Annual sinking fund	$\dfrac{s}{(1 + s)^n - 1}$
Also			
Years' purchase dual rate adjusted for tax			$\dfrac{1}{i + \left[\dfrac{s}{(1 + s)^n - 1} \times \dfrac{1}{(1 - t)} \right]}$
Years' purchase in perpetuity			$\dfrac{1}{i}$
Years' purchase of reversion to a perpetuity			$\dfrac{1}{i(1 + i)^n}$

Note: the term 'years' purchase' is commonly used in preference to the more descriptive 'present value of £1'.

Key: i, interest (shown as a decimal fraction); n, years; s, sinking fund rate (where different from remunerative rate); t, tax rate (shown as a decimal fraction).

The present value of £1

The single amount to be invested now to accumulate to £1 in 'n' years at 'i' inter est. It is self-evident that the prospect of £1 at some future date is worth less than £1 available today. The precise amount of the present value will be determined by the period of deferment and the rate of interest.

Example 4.1

A legacy of £5000 is payable in four years' time when the recipient reaches the age of 21. Calculate the present value of the sum where the appropriate rate of interest is 6 per cent.

Legacy payable in 4 years' time	£5 000.00
PV £1 in 4 years @ 6.5%	0.7921
Current value of legacy	£3 960.47

Step-by-step workings

Legacy due in 4 years' time	£5 000.00
PV £1 in 1 year @ 6.5%	0.9434
Value in 3 years' time	£4 716.98
PV £1 in 1 year @ 6.5%	0.9434
Value in 2 years' time	£4 449.98
PV £1 in 1 year @ 6.5%	0.9434
Value in 1 year's time	£4 198.11
PV £1 in 1 year @ 6.5%	0.9434
Current value	£3 950.50

(Slight difference due to rounding.)

Commentary

The value of the legacy today is appreciably less than the amount receivable in four years' time. The step-by-step workings are shown, a laborious process when the period is longer than a few years.

The amount of £1

The formula is designed to show the total amount of compound interest added to the initial and only deposit of £1 for a specified time and at a particular rate of interest. It assumes that interest is added at the end of each year and is left in the account to itself attract interest in subsequent years (the principle of compound interest).

Example 4.2

A retailer has entered into a new lease, which includes a commitment to spend £20 000 on internal rearrangements to the premises. A loan has been arranged at an interest rate of 7.5 per cent to be repaid at the end of three years. Calculate the amount due at that time.

Loan for internal rearrangements	£20 000.00
Amount of £1 in 3 years @ 7.5%	1.2423
Repayment due in 3 years	£24 846.00

Check, showing year-by-year calculation:

Loan	£20 000.00
Amount of £1 in 1 year @ 7.5% (i.e. × 0.075)	£ 1 500.00
	£21 500.00
Amount of £1 in 1 year @ 7.5%	£ 1 612.50
	£23 112.50
Amount of £1 in 1 year @ 7.5%	£ 1 733.44
Amount for repayment in 3 years	£24 845.94

Commentary

At the rate of interest charged, the total cost of borrowing over the period is quite substantial as will be seen. As a check, the amount due at the end of the period is shown year by year, a process that would be impracticable for long periods.

Years' purchase (or the present value of £1 per annum)

The present value of the right to receive £1 at the end of each year for 'n' years at 'i' compound interest. This formula is used to find the capital value of the right to receive a stream of income, typically but not necessarily a rent. The same result could be found by adding together a series of present values (PVs) of £1, which is shown as an alternative. Again, this would be practicable only for short periods.

Example 4.3

Your client owns a freehold office building recently let at a market rent of £30 000 per annum. The rent will be reviewed at the end of five years. You are asked to indicate the present value of the first five years' income at a discount rate of 6.5 per cent.

Rent of office building per annum	£ 30 000.00
YP £1, 5 years @ 6.5%	4.1557
Value of first 5 years' rent	£124 671.00

Rent per annum		£ 30 000.00
PV of £1 in 1 year @ 6.5%	0.93897	
PV of £1 in 1 year @ 6.5%	0.88166	
PV of £1 in 1 year @ 6.5%	0.82785	
PV of £1 in 1 year @ 6.5%	0.77732	
PV of £1 in 1 year @ 6.5%	0.72988	4.1557
Value of first 5 years' rent		£124 670.40

Replacement of capital by sinking fund at same rate

Rent		£ 30 000.00
Return on capital value @ 6.5%		£ 8 103.62
		£ 21 896.38
Amount of £1 per annum for 5 years @ 6.5%		5.6936
		£124 669.23

Commentary

A payment of £30 000 a year will be received at the end of each of the next five years. The capital value will depend on the discounting rate used, in this case, 6.5 per cent. The calculation will reflect the receipt of a total of £150 000 over the next five years when the first payment is deferred for one year, the next for two years and so on.

A feature of compound interest is that it not only services the capital but also leaves a balance, which, if reinvested at the same rate, replaces the original capital sum at the end of the term, as demonstrated in this example.

Amount of £1 per annum

The formula shows the amount to which £1 invested at the end of each year would accumulate at a given rate of interest.

Example 4.4

Janet intends to save £3000 each year for the next three years. Calculate the value at the end of the period if the savings are invested in a tax-free ISA account at an interest rate of 5 per cent.

Savings each year	£3 000.00
Amount of £1 per annum, 3 years @ 5%	3.1525
Amount available at end of 3 years	£9 457.50

Check with yearly calculation:

Amount invested at end of year 1	£3 000.00
Interest added at end of year 2	£ 150.00
Balance carried forward	£3 150.00
Amount invested at end of year 2	£3 000.00
Balance carried forward	£6 150.00
Interest added at end of year 3	£ 307.50
Balance carried forward	£6 457.50
Amount invested at end of year 3	£3 000.00
Total return at end of 3 years	£9 457.50

Commentary

A total of £9000 will be saved over the period and interest will be added to the increasing sum. Note that amounts are invested at the *end* of each year. The progressive accumulation of capital and interest is shown.

Annual sinking fund

The annual premium required to be invested at the end of each year to accumulate to £1 at a specified compound rate of interest over a period of years. One of its principal uses is to provide for the replacement of capital in investments that have a limited life, referred to as wasting assets, in which case it will be linked to the years' purchase formula.

A feature of compound interest is that the interest payments include not only interest on the capital sum invested, but an additional amount, which, if reinvested at the same rate of interest, will return the original amount expended at the end of the period.

Example 4.5

The tenant of a small flat has agreed to purchase it in five years' time when a deposit of £25 000 will be required. How much should be invested each year if a rate of interest of 5.25 per cent can be obtained from a five-year fixed interest bond underwritten by the government?

Deposit required in 5 years' time	£25 000.00
Annual sinking fund, 5 years @ 5.25%	0.1819
	£ 4 547.50

Commentary

This calculation will enable the tenant to save an adequate amount each year to provide the necessary deposit when compound interest is taken into account. The requirement that a sinking fund should be risk-free is satisfied by the guarantee accompanying the bond. Tax on interest has been ignored.

The annuity purchased by £1

The formula envisages investing a single amount now to provide an income for a specified period commencing at the end of the first year. The capital and interest accruing will be returned over the life of the annuity. The whole of the fund will have been used by the end of the period.

Example 4.6

Your client is taking early retirement and wishes to purchase an annuity to supplement investment income pending receipt of a pension in 10 years' time. Advise on the annuity available from a capital investment of £15 000 at a discount rate of 5.5 per cent.

Sum available to purchase annuity	£15 000.00
Annuity, 10 years @ 5.5%	0.1327
Annuity	£ 1 990.50

Commentary

The whole of the fund and the interest earned will be applied in making annual payments over the period of 10 years. There will be no terminal value.

Other formulae

Formulae required for other eventualities are described below.

Years' purchase in perpetuity (the present value of £1 per annum in perpetuity)

If it can be assumed that the income will be received indefinitely on a yearly in arrears basis, the formula is of the simplest form. The years' purchase is found by dividing the decimal equivalent of the appropriate yield into unity.

Example 4.7

Calculate the capital value of the right to an income of £5000 per annum in perpe-tuity where the appropriate return by reference to the sales of other similar properties is 8 per cent.

Annual rent	£ 5 000.00
Years' purchase in perpetuity @ 8%	12.50
Capital value	£62 500.00

Commentary

The return of £5000 per annum on an investment of £62 500 can be checked to show that the interest rate is indeed 8 per cent.

Years' purchase (YP) in perpetuity deferred 'n' years

If receipt of payments is deferred for an initial period, the effect can be calculated in one of two ways, both of which give the same result:

- YP in perpetuity minus YP for initial period of delay
- YP in perpetuity × present value of £1 for initial period of delay.

The following example will be used to demonstrate both approaches.

Example 4.8

Assume the right to receive £1000 per annum in perpetuity after the expiration of three years from now. The current interest rate may be taken as 5 per cent.

Income in perpetuity commencing in 3 years' time		£ 1 000.00
YP perpetuity @ 5%	20.00	
Less YP £1, 3 years @ 5%	2.7232	17.2768
Capital value		£17 276.80

Also

Income in perpetuity commencing in 3 years time		£ 1 000.00
YP perpetuity @ 5%	20.00	
× PV £1, 3 years @ 5%	0.8638	17.2760
Capital value		£17 276.00

(Slight differences due to rounding of components.)

Commentary

The capital value is modified to reflect that no income is received during the first three years. Printed tables are necessarily limited because of the infinite number of permutations of yield and deferral.

Years' purchase, dual rate, adjusted for tax on sinking fund element

Example 4.9

A client wishes to purchase a leasehold interest that has an unexpired term of 40 years and a current rent of £6000 per annum. The head rent is fixed at £250 per annum. It is anticipated that investors would expect a yield of 9 per cent.

Rent received per annum	£ 6000.00
Less head rent	£ 250.00
Net income	£ 5750.00
YP 40 years @ 9% and 4% adjusted for tax @ 25%	9.6125
	£55271.88
Allocation of income	
Rent received	£ 5750.00
Less interest @ 9%	£ 4974.48
	£ 775.52
Deduct tax @ 25%	£ 193.88
Available for sinking fund	£ 581.64
Check amount of £1 p.a., 40 years @ 4%	95.0255
	£55270.63

(Slight difference due to rounding.)

Commentary

The remaining lease term of 40 years is regarded as a wasting asset requiring replacement of the original capital at the end of the term. The years' purchase formula is modified to include a sinking fund taken out at a safe and therefore low rate of interest to provide for the original capital to be replaced. As the premiums for the sinking fund will be payable from taxed income, the effect on tax must also be provided for. Selection of the tax rate creates some difficulty but should reflect the rate paid by companies or individuals likely to engage in such activities. It should be noted that some investors, such as charities, enjoy certain tax benefits and would be able to consider the provision of a sinking fund from gross income.

The provision of a sinking fund has a limited effect on the capital value, given the remaining term. As a comparison, an income in perpetuity at the same yield would be worth approximately £8000 more.

There is no evidence that investors do take out a sinking fund, but its inclusion in the years' purchase table enables a comparison with freehold market yields.

The analysis included in the solution demonstrates the allocation of income between return on investment and provision for sinking fund after allowing for tax on that portion of the income.

Compound interest provides sufficient income to service the interest on capital and to provide a sinking fund to replace the original capital sum, but only where the same yield is used for both purposes.

Any income from a wasting asset is expected to be sufficient to provide not only a return *on* capital but also a return *of* capital. The dual rate table is designed to reflect the difference in valuation terms between perpetual and lesser terms.

Receipt of rent quarterly in advance

The majority of modern leases provide for the payment of rent quarterly in advance. It follows that transactions should be analysed on this basis and valuations undertaken with that direct evidence of yields.

But the use of the annual in arrears formula is not only well established practice, it is also easy to use, the calculation of any years' purchase in perpetuity being no more than a simple division of 100 by the yield. That relationship no longer holds good for a quarterly payment. The years' purchase on a quarterly in advance basis involves looking up tables, having a computer program or making a set of calculations for each year's purchase required.

The formula for calculating the years' purchase for an income received quarterly in advance is:

$$\frac{1 - (1 + i)^{-n}}{4 \left[1 - \dfrac{1}{4 \sqrt[4]{(1 + i)}} \right]}$$

Formula 4.2 Years' purchase for rent payable quarterly in advance

It must be emphasized that whichever approach is used to make the analysis must also be used in any valuation, using the information gained from that transaction.

The years' purchase of a rent payable annually in arrears is lower than the comparable years' purchase for a rent payable quarterly in advance. An example will show the difference.

Example 4.10

Value an income of £1000 per annum, receivable for the next ten years on the single rate basis, yielding 8 per cent:
- using the years' purchase annually in arrears
- using the years' purchase quarterly in advance.

	10 years	Perpetuity
• Payments received annually in arrears	6.7101	12.5
• Payments received quarterly in advance	7.0424	13.1190

Years' purchase for life or lives

Life interests are indeterminate terms in freehold or leasehold property. The person having the benefit of a life interest is referred to as the life tenant or tenant for life. There may be a single life tenant, joint life tenants, or a tenancy for the longer of two or more lives. Such interests are rare, surviving from the time when many estates were settled to maintain them in family ownership. When the head of the family predeceased his wife, it was a common arrangement to leave the estate to her for the duration of her life with remainder to the heir or heirs.

The process of valuation is complicated and precarious, the more so because there is uncertainty about the length of the interest. English Life tables compiled by government statisticians from census information and recorded mortality of a sample of 100 000 people enable probability factors to be calculated. These are then used with a years' purchase figure to reflect the uncertainty. Any individual may live for a longer or shorter time than the statistical result and an individual valuation is therefore highly speculative. The interest belongs in the class referred to as wasting assets.

It is likely that the majority of valuers will complete their professional lives without ever performing a valuation of a life interest.

CONTEMPORARY APPROACHES

The mathematical processes associated with the contemporary approach of discounted cashflow and its derivatives are presented in Chapter 7.

5 The determinants of value

INTRODUCTION

The purpose of an investment valuation is to provide an opinion on the capital value of the right to receive regular streams of income for a definite or indefinite period.

The valuation may be prepared on behalf of the owner in the form of advice preparatory to marketing, for a prospective investor, or for a third party contemplating the granting of a loan secured on the property.

The majority of valuations are prepared on the basis of market value, which is defined as:

> the estimated amount for which a property should exchange on the date of valuation between a willing buyer and a willing seller in an arm's length transaction after proper marketing wherein the parties had each acted knowledgeably, prudently and without compulsion.
>
> © *RICS Valuation Standards (The Red Book) 6th edition*, published 2007
> (effective from 1 January 2008)

On receiving instructions, the valuer should ascertain the purpose of the valuation as it may affect the result or more particularly the way in which it is reported. If the valuer is instructed to report to an investor on the worth or some other 'non-market' basis, the report should make clear that the valuation has been prepared having regard to particular instructions and that the outcome does not necessarily represent the valuer's view of the market value of the premises.

Any consideration of capital value involves the collection of many details leading to a view about the quality of the investment as part of the process of preparing a valuation.

It is also necessary to determine what is being valued. Each separate property is unique: even when it is indistinguishable in form from an adjoining property, it occupies a different site, the location of which may be of great importance in the consideration of value. The real importance usually lies in the quality and quantity of the particular income, in the limitations of legal title and in the constraints imposed by law, for example on the use of the premises and on the landlord–tenant relationship.

The initial part of the process relies on the skill and judgement of the valuer in identifying as much information as possible. The valuer is then in a position to finalize the valuation.

Apart from the amount of rent, the specific provisions of the lease and the physical attributes of the property, most aspects are open to judgement and will inevitably attract varying levels of interpretation and significance from one valuer to another. Such judgements may overlap the quantitative information previously referred to. For example, an expensively constructed building is not necessarily the best building from a tenant's point of view. The collective view of the market will be what determines the level of rental value, and while the form and content of the lease is known, any interpretation of its effect may be open to question. Similarly, the standing of a particular company as a tenant will be a factor open to different views by different valuers.

The consideration of capital value involves the collection of a range of information, which may be discovered from a physical inspection, perusal of leases and other legal documents, and confirmation of the use class under planning legislation including any restrictions or limitations imposed, payments made by or to the owner for wayleaves, licences and other grants of rights. Where there is any doubt, the valuer should seek clarification. For example, it is unlikely that title deeds will be available for inspection, in which case information should be obtained from the owner's solicitors before proceeding. The next step is the quantification of income, either actual or estimated, together with the amount of any outgoings.

Having obtained whatever information is available on recent sales and lettings, the valuer will analyse that information and decide whether it is relevant and if so to what extent and how it may contribute to the valuation in preparation. Differences in location, age of building, permitted use, lease terms, standing of tenants and future prospects will all be relevant. As already mentioned, while the form and content of the lease is known, the interpretation of its effect may be open to discussion.

Opinions about rental income will vary. Growth is a function both of the longer term future of the building itself and of the state of the economy, a complex consideration in which any opinion is likely to rely heavily on the evidence of past performance for want of better information.

But even here there is room for opinion: the current rental value and investment yield will be influenced by the general availability of the class of property under review, the demand for it, and even on proposals for future developments. Investigations and enquiries to obtain the information required should be exhaustive since the soundness of the valuation will be determined by the thoroughness of the collection and interpretation of data.

The status of a property in the investment market is an amalgam of these considerations, which rely heavily on the interpretational judgement of the valuer. Market value cannot be determined without a good deal of investigation, both of the facts surrounding the particular building and of the circumstances in which similar properties were sold. With regard to information on other transactions, it is worth pointing out that the valuer is rarely embarrassed by an excess of information and so would be well advised to examine any recent transactions to judge their relevance.

The following extract from a decision given by Megarry J. (as he then was) provides a succinct account of the way in which any competent and careful valuer would approach the question of valuation:

> In building up his opinions about values, he will no doubt have learned much from transactions in which he has himself been engaged, and of which he could give first-hand evidence. But he will also have learned much from many other sources, including much of which he could give no first-hand evidence. Textbooks, journals, reports of auctions and other dealings, and information obtained from his professional brethren and others, some related to particular transactions and some more general and indefinite, will all have contributed their share. Doubtless much, or most, of this will be accurate, though some will not; and even what is accurate so far as it goes may be incomplete, in that nothing may have been said of some special element which affects value. Nevertheless, the opinion that the expert expresses is none the worse because it is in part derived from the matters of which he could give no direct evidence. Even if some of the extraneous information which he acquires in this way is inaccurate or incomplete, the errors and omissions will often tend to cancel each other out; and the valuer, after all, is an expert in this field, so that the less reliable the knowledge that he has about the details of some reported transaction, the more his experience will tell him that he should be ready to make some discount from the weight that he gives it in contributing to his overall sense of values. Some aberrant transactions may stand so far out of line that he will give them little or no weight. No question of giving hearsay evidence arises in such cases; the witness states his opinion from his general experience.
>
> *English Exporters (London) Ltd v. Eldonwall Ltd (1973)*

As indicated above, the quality of an investment is a subtle combination of a number of distinct features, the precise effect of each on the particular investment being a matter of mature judgement. Various aspects of investment in property are now discussed in some detail.

THE PHYSICAL ENTITY

The building

The investor will prefer a building that is recently constructed of first class materials, of a design that has aesthetic appeal, and the use of which is adaptable to possible future changes in demand. Further, the site should be in an appropriate location for its particular use.

If repairs become necessary, it should be possible to carry them out speedily, without undue expense, and without heavy reliance on specialist proprietary replacement parts, which may become difficult to obtain over the life of the building. This appears to raise questions about many 'systems' buildings, very popular in the 1960s but which often now require major rehabilitation to enable their continued use.

The frequency and cost of repairs is important, even when the tenant is responsible for such work. Not only will a building known to be expensive to maintain be more difficult to let or re-let, but any prospective occupier will examine the likely burden of costs of repair when considering the level of rent to be negotiated – the tenant is more concerned with total occupation costs than the component parts of such costs. The level of expenditure will also be a factor in rent review negotiations.

The foregoing specification is one of perfection that few buildings will attain; the point is that any departure from the ideal is a factor to be taken into account when considering the question of value. A few minor shortcomings may be acceptable but several more important deficiencies are likely to narrow the potential market for both tenants and purchasers, with a consequent effect on rental and capital values.

The site

The site and its location are important. The valuer will wish to judge the suitability of the site for the building erected on it. Considerations will include whether it is large enough and caters for any particular requirements associated with its use, and whether it has adequate off-site car parking provision and suitable loading and unloading facilities. The relevance of the location to its use will always be important and, under some circumstances, critical.

The location of shop premises is limited to particular, relatively small, central areas and a difference of a few metres can make a substantial difference in value. Office buildings tend to develop in clusters; one part of a town or one street in a town often gains a reputation as the location for the majority of, for example, law firms or accountants, which will act as a magnet to others in the same or associated professions. Public transport or accessible on-site car parking provision is desirable, failing which convenient public car parking provision should be expected. Staff recruitment may be helped by the proximity of shopping, public transport and car parking provision.

Factories need good road access and proximity to a sufficient workforce, although their precise location is rarely of prime importance. Retail warehouses rely on a prominent main road position with good access and ample parking. Retailers and shoppers show a preference for a purpose-built retail park housing a number of suppliers rather than the earlier, single-site development.

The precise form of the original planning permission needs to be considered; many of the early retail warehouses have been able to add a mezzanine floor within the original permission. Later permissions tend to state the maximum retail area to be provided or prohibit the addition of an upper floor, even within the existing structure.

CONTRACTUAL MATTERS – LEASE TERMS

The lease is a contract between landlord and tenant that controls their relationship throughout its term and in some cases beyond it. A badly drawn lease is likely to be all too apparent in the case of any disagreement. The various provisions of a satisfactory lease will exert considerable influence over rental and capital values and are worthy of careful investigation.

Landlords aspire to a modern institutional type of lease, which may be described as a lease for a term of from 15 to 25 years with provision for upward only rent review at intervals not exceeding five years and with the tenant responsible for the execution of all repair and maintenance work and for the cost of insuring the premises including cover for loss of rent in the event of damage to or loss of the premises. Not every property is suitable for such a lease and the final terms will normally be the outcome of negotiations between the parties. The owner of a large development may insist on the use of a standard lease throughout, in which case the prospective tenant is faced with the decision either to accept it or to decline to take a unit on that development.

More recently, there has been some reaction against the unfairness of 'upward only' review clauses, under which the tenant undertakes to maintain the rent at its current level at least, even if the rental value has fallen. Against the background of government suggestions that the practice should be outlawed by legislation, representatives of landlords, tenants and government have combined to produce the *Code for Leasing Business Premises in England and Wales 2007*. The code applies specifically to new leases and sets out the guiding principles for landlords and occupiers. The emphasis is on clarity and fairness. In this context, it is suggested that alternatives to the upward only review could be offered, a simple approach to break clauses, proper protection and interest payments on any rent deposits and a simple process for assignment. The code is voluntary.

Rent

The valuer will seek information about the current rent payable, the frequency with which it is paid, when it is due, whether it is payable in advance or in arrears and what sanctions are available when payment is late (modern leases often contain provisions for a fairly punitive rate of interest of 3–4 per cent above bank base rate to be levied on overdue payments).

The other terms of the lease will be examined to determine the extent to which the landlord remains liable for any of the outgoings. In the ideal case (from the investor's point of view), the tenant will be responsible for the costs of all repair, maintenance and insurance. If the landlord remains liable for some outgoings, he or she has an uncertain liability. The market is likely to reflect this disadvantage and uncertainty by expecting a higher yield than that required on a full repairing lease.

If the rent being paid is less than the rental value, the valuer will need to analyse the available evidence to enable the market rent to be assessed.

There will be regular charges for management if an agent is employed, and further legal and other professional charges may be incurred if disputes take place, if the rent is reviewed, or if the property is re-let, either to the existing tenant or to a new tenant.

Rental value

Any valuation must take account of net income. If it is anticipated that the rent payable will alter, this should also be reflected. Should the landlord retain any substantial responsibility for repairs or insurance, the effect of these expenses should be estimated in arriving at the net rent.

Estimation of the current rental value involves a judgement of the market, and adds a further dimension to the process of valuation. It should be noted that in the traditional approach to valuation, no attempt is made to project the rental value to the time when it first becomes payable. The valuer is simply required to estimate the current rental value at the time of the valuation and then to take it into account in the valuation from the earliest time that it may be imposed.

Rental value is estimated by an analysis of whatever relevant information of rent levels is available to the valuer, who should have a good knowledge of the market but may not always be aware of all the transactions that have taken place.

Once a prevailing pattern of rental values has been established, it will be used to reach a view of the rental value of the particular unit under consideration, bearing in mind that not all the evidence may be relevant or relate to truly comparable properties. For example, if the unit is appreciably larger or smaller than the properties from which the basic information is derived, the valuer's judgement may suggest that there is a need for some adjustment. Regard must be had to age or other physical attributes and the legal framework within which the property is let. When the current tenant is likely to seek a new lease, regard must be had to the effect of any relevant landlord and tenant legislation, particularly exclusion of the value of any relevant improvements made by the tenant. The likelihood or otherwise of any unsatisfactory lease terms being changed in negotiations between the parties or by the court must be taken into account, including the effect on rent of any unusual review pattern or onerous repairing obligation likely to be replicated in a new lease.

Rent reviews

It is important in maintaining the value of the investment that the rent is maintained at a realistic level. It is usual in modern leases to provide for regular rent reviews. Most landlords would not wish a rent to be fixed for a period of longer than five years. The form of the review will depend on the arrangements between the parties as contained in the lease. The most common type is the upward only review, but there are others, all of which are described briefly below.

Upward only review

The principle is that the rent will be reviewed at predetermined intervals and a new rent assessed. The rent to be paid following the review will be the greater of either the rent passing immediately before the rent review or the full rental value on the date of review, whichever is the greater. This form ensures that, even should the rental value go down, the minimum rent will be the rent passing during the previous period.

The introduction of a voluntary code following government pressure has been referred to earlier.

Predetermined rent review

Sometimes known as escalator rents, the landlord and tenant agree future increases at the beginning of the lease. The advantage to both parties is the certainty of future payments and the cost-saving because there is no necessity to involve surveyors or to become involved in expensive referrals to court or arbitration. The drawback is the absence of a market rent, which may benefit the landlord or the tenant.

Index-linked rent review

The parties agree at the outset that the rent will vary according to an index such as the Retail Price Index (RPI). It may change on a yearly basis or at intervals, as with other forms of rent review. The problem with external indices is that their basis may change or that they may not reflect aspects that have an influence on rental values.

Upwards or downwards rent review

The rent is reviewed at intervals but may go up or down according to values current at the time of the review. This arrangement ensures that the rent, often one of the principal overheads, does not move ahead out of line with commercial expectations.

Turnover rents

Turnover rents typically consist of a fixed base rent together with a percentage of the turnover once it meets an agreed level. The arrangement in effect provides for an annual review, although it could be upwards or downwards depending on the profitability of trading. More information on turnover rents is given in Chapter 10.

INTERPRETATION OF LEASE TERMS

Most leases are unique and no assumptions should be made about their contents. The valuer has no short cut: it is necessary to read the whole of the lease in the context of the statutory background and in relation to the formidable body of case law that has grown up in recent years. It would be appropriate to seek advice about the effect of any provision that is of concern to the valuer.

It is not proposed to deal in detail here with the law, but it is pertinent to highlight those areas that affect value, especially where there has been considerable court activity that is likely to influence market value.

In addition to the repairing covenant and the provisions for rent review, the landlord will seek to exercise some control over the use of the property and to limit the tenant's ability to carry out work or assign or sublet his interest to another.

Specification of the user clause should receive careful consideration to avoid any lowering effect on the rental value that may result from a strict or narrow user provision. In one case, the requirement to use offices only for the work of consulting engineers led to a restriction of the rent imposed on review (*Plinth Property Investments Ltd v. Mott Hay & Anderson (1978)*). The landlord of business premises where such a user clause has an adverse effect on rent may seek to change the provision on a renewal under the 1954 Act but is unlikely to be successful without the agreement of the tenant (*Charles Clements (London) Ltd v. Rank City Wall Ltd (1978)*).

Where the tenant of business premises wishes to carry out improvements, the landlord can offer to do the work in exchange for a reasonable additional rent, but otherwise will find it difficult to prevent the tenant from undertaking the work. Where the tenant undertakes the work, the provisions of the Landlord and Tenant Act 1954 ensure that, on renewal, the effect of the improvements is disregarded in setting the new rent. However, the situation on review will depend on the provisions contained in the lease. In one case, an unfortunate tenant found himself paying rent for improvements that he had recently financed as an addition to a rebuilding following a fire (*Ponsford v. HMS Aerosol (1979)*).

The wish of a tenant to assign the lease is often opposed by a landlord who may employ delaying tactics. Recent legislation (Landlord and Tenant Act 1988) has shifted the balance, enabling the tenant to proceed when the landlord has not responded to any request within a reasonable time. The landlord will often prohibit subletting on the grounds of deterioration in value, possible extra costs of management and the risk of additional health and safety requirements being imposed.

While the repairing covenant under a full repairing lease places considerable burdens on the tenant, including the responsibility to renew parts of the structure where necessary, it does not extend to completely renewing a failed building.

It has been described as:

> always a question of degree whether that which the tenant is being asked to do can properly be described as repair, or whether on the contrary it would involve giving back to the landlord a wholly different thing from that which he demised. In deciding this question, the proportion which the cost of the disputed work bears to the value or cost of the whole premises, may sometimes be helpful as a guide.
>
> *Ravenseft Properties Ltd v. Davstone (Holdings) Ltd (1979)*

In this case, the tenant was held liable to pay for a new method of fixing concrete cladding and to provide expansion joints not previously provided since the work proposed was the only way of remedying the defect. But in another case, where a house suffered from severe condensation problems, it was held that there was no evidence of lack of repair. The cause of the complaint was a design fault, remedying which would amount to an improvement (*Quick v. Taff Ely Borough Council (1962)*).

Provisions for rent review have become more sophisticated and complicated in recent years, alongside the realization by landlords that regular increases in rent are an integral part of the value of the property. Many leases lay down stringent provisions in the machinery for undertaking reviews, neglect of which may seriously disadvantage the tenant. Some of the provisions create unreal situations where the rent is to be assessed in hypothetical circumstances unrelated to the market. When, usually too late, the effect of the provision is realized, there is often tension in the landlord–tenant relationship.

It is a source of some surprise that tenants continue to agree upward only rent reviews that may result in the imposition of reviewed rents above the level of the market. So far there have been few cases where this restriction has made any difference to the rent level, although in some sections of the office market tenants wishing to vacate during the course of the term of the lease have found it necessary to offer a lump sum (a reverse premium) to attract an occupier or to persuade an owner to accept a surrender of the lease. From an investor's point of view, rents in excess of the intrinsic rental value add a further risk dimension to property investment.

The landlord will expect to be consulted before any alterations are carried out. In many cases, work that the tenant proposes to undertake will be of long-term benefit to both parties, although specialist adaptations may reduce the letting value in the market. The landlord's ability to resist improvements proposed by the tenant is much constrained by legislation. Unless the landlord offers to do the work, the tenant may apply to the court for permission if the landlord continues to resist the tenant's application.

Whenever a dispute arises, the result turns upon the facts, which are unlikely to be the same in two different cases, leading to uncertainty about the outcome of any case going to court. Litigation is expensive and it may be that an apparent flaw will be exploited by one party because the other party is unable or unwilling to risk a court case. On the other hand, investors with large land holdings may take what in themselves are quite trivial cases in order to establish, or at least to clarify, a particular principle.

The valuer needs sufficient understanding of the law of landlord and tenant to be aware of the significance of particular provisions, especially where they have been the subject of interpretation by decided cases.

The valuer is not trained in legal interpretation, and if he or she is uncertain about the precise meaning of a clause that is likely to have an effect on the value, expert legal opinion and advice should be sought.

THE TENANT'S 'COVENANT'

The financial standing of the tenant is important for trouble-free financial relationships. If the tenant is a major commercial concern, perhaps a public limited company with shares quoted on the stock exchange, the rent is likely to be more secure than if the tenant is a recently formed expanding firm experiencing cash flow problems and without ready access to outside funds.

The advantages of payment of rent by direct debit on the due date is a feature that can be appreciated fully only by those who have experienced the frustration and time involvemed in trying to collect rent from unsatisfactory tenants. The clear implication is that tenants should be selected with great care, and it is part of the valuer's concern to assess the quality of a tenant's covenant to pay rent and perform the other obligations under the lease.

A good tenant will not only pay the rent on time, but may be more likely to honour repairing and other obligations. The cost of management is thus minimized and the attraction of this type of tenant is such that prospective purchasers may be prepared to recognize the benefits in the acceptance of a slightly lower yield.

New developments are a particular example of the benefit of a known tenant. If the developer of a shopping centre is able to let space to a major company as an 'anchor' tenant, confidence is created that often results in the attraction of other tenants. The large retailing groups are aware of their power and use it to negotiate attractive initial terms in such circumstances.

Any investor is first concerned to know the minimum net income to be generated by a prospective investment. Because of the nature of real estate, that exercise is in itself of a different complexity from that in the case of government stock or shares, where the dividend is received free of outgoings and, in most cases, after tax has been deducted from the payment.

OUTGOINGS

Landlords endeavour to agree a letting on full repairing and insuring terms under which the tenant accepts responsibility for all costs of repair and maintenance and for the insurance on a comprehensive basis of the building, its plant and fixed machinery.

When the property is let on such terms, the valuer will be concerned only with the costs of management incurred by the landlord. In other cases, an estimate of the likely annual costs of maintenance or other expenditure for which the landlord remains liable will be necessary, together with an estimate of the cost of any accrued repairs or maintenance.

The costs of management include not only any annual management fee but also annualized amounts for negotiations of rent reviews and lease renewals, and for other services such as periodic assessments for insurance cover where such work is not included in the basic management fee and is not recoverable by the landlord.

Approximate estimates of outgoings are sometimes achieved by taking percentages of the rental value, based on experience of managing similar properties.

Landlords regard the full repairing and insuring lease (where the tenant is responsible for all outgoings) as an important factor for two reasons. First, if the landlord retains responsibility for an outgoing, the detailed management involved in identifying and organizing any work necessary is expensive and erodes the rental income.

Second, the annual costs over the length of a review interval tend to increase, reducing the real value of the fixed rent. The market will tend to reflect its preference by seeking a higher yield where the landlord retains substantial repairing obligations. Alternatively, it may take the more drastic action of not considering the acquisition of investments where there is significant landlord involvement in dealing with outgoings.

The landlord will wish to be satisfied that the tenant is financially able to meet the probable demands of the covenant.

Tenants would be well advised to consider the condition of the building carefully before taking on responsibility for such onerous terms. In particular, a building in the later stages of its usefulness may not be suitable, from the tenant's point of view, for agreeing to the terms of a full repairing lease.

YIELD

The yield required by the market may be analysed from recent transactions involving the sale of similar investments. The question of comparability requires careful consideration. Unless there is similarity in the various determinants of value discussed above between the property to be valued and the evidence obtained from recent transactions, there is potential for misleading conclusions.

The size of the property, its age, the quality of tenant, the form of the lease (particularly its provisions relating to repairing obligations and the interval between rent reviews) and the period before the next rent increase can be implemented, all present opportunities for differences. The conventional approach to valuation, soon to be discussed, reflects all these similarities and differences in a yield analysed from known transactions and adjusted at the discretion of the valuer. The yield is then used to calculate a year's purchase (similar in effect to a price:earnings ratio).

However, rent is only one aspect of the investment return. As the rent increases, so the investor may expect the capital value to rise also (unless there is a change in the investment market whereby investors look for an increase in the yield). When the market's fixed interest requirement can be ascertained, it is possible to calculate the income growth inferred by analysing the all-risk yield.

SERVICE CHARGES

Commercial premises

Under the provisions of many leases, the landlord maintains closer control of the work for which the tenant is responsible by managing that work and then charging the tenant for the work and a management fee. The procedure is particularly useful, and probably essential, where the premises are let to a number of tenants.

Services provided to tenants of business premises are administered and charged in accordance with the contractual arrangements agreed between the parties as contained in the lease.

Residential tenants were thought to need more protection from the sometimes harsh contractual terms, and statute has interceded to regulate the relationship, as described later.

The service charges on commercial premises are determined by the provisions of the lease. There is no overriding legislation, as with residential premises.

Lease stipulations for the range, provision, apportionment and collection of service charges become very complex in the case of an office block in multiple occupation or a retail shopping development. It is important for the arrangements to be clear and the implementation efficient. The modern lease usually lists a comprehensive range of services that are controlled by the landlord, with further unidentified services to be provided if found to be necessary.

The landlord will normally enter into a number of contracts for items such as cleaning and refuse collection, and should ensure that the terms of the contract are complied with. Certain services, such as security or landscape maintenance, may be delivered directly through staff employed by the owner.

The lease will set out the method of cost apportionment. Each tenant's share of the total cost of services may be related to the floor area, a weighted floor area to reflect different requirements, rating assessments, the RPI or the rent paid. A tie in to any datum point outside the control of the landlord (such as rating assessment or RPI) is best avoided as changes may have unexpected and unwanted effects.

The collection of service charges usually takes place in two parts. An estimate is produced of the total costs for the year on the basis of which the landlord collects advance quarterly payments, the balance being payable once the actual costs are known and certified.

It is important for estimated costs to be paid in advance, otherwise the landlord will need to have funds available to finance expenditure until tenants' payments are received. In general, interest charges will not be recoverable as part of a service charge unless specifically provided for in the lease. There are many unsatisfactory service charge provisions in existence, with the result that the landlord may be unable to collect the whole expenditure, or the cost of the working capital required to deliver the service has to be subsidized.

Value Added Tax (VAT)

Rent payable under a lease will, as a general rule, be exempt from VAT. A landlord may opt to tax, turning an exempt supply into a standard rating one.

If a landlord exercises this option, rental income from certain property that would normally be exempt will become standard rated. A landlord would exercise such an option to allow the recovery of input tax that would otherwise be lost.

Once the landlord has opted to tax, all supplies in connection with that particular interest in the property must be standard rated.

The landlord should consider the position of a tenant or a potential purchaser when considering whether to opt to tax. If the tenant or purchaser cannot recover the VAT charged, opting to tax may not be a good commercial decision.

When an investor has opted to tax and is able to reclaim VAT on certain payments made, the effect of VAT may be ignored. If the company or group has not opted to tax or if a modified rate is charged, the effect on the transaction should be considered and any net cost included in the calculations.

BUSINESS TENANCIES

The impact of statutory intervention

Most investments in property involve occupation for business purposes. The fact that there is a thriving investment market in business properties enables occupiers to have the confidence to become tenants rather than tie up capital in expensive premises.

Since 1954, business tenants have enjoyed considerable security of tenure, and in cases where possession is gained against the wishes of the tenant, statutory compensation is normally available. Both events have implications for the valuation process.

There is a public interest case for imposing limitations on a landlord's ability to obtain possession of business premises when the tenant may have invested considerable time, skill and funds in creating goodwill (where it cannot readily be translated or compensated).

The first attempt to give tenants some legislative protection was the Landlord and Tenant Act 1927, which failed to achieve its objectives and was repealed when replaced by the 1954 Act, except for those parts concerned with tenants' improvements.

Changes to main provisions

The Landlord and Tenant Act 1954 (as modified by the Law of Property Act 1969) was and remains the principal Act relating to business premises. But important changes were introduced in 2004 by the Regulatory Reform (Business Tenancies) Order 2003 (SI 2003/3096), which amended parts of the 1954 Act. The 2003 Order reflects to a large extent changes in business needs since 1954 and seeks to give both parties greater freedom to negotiate tenancies that suit them. In particular, it recognized that there were circumstances in which the parties were prepared to enter into a contract without the protection afforded by sections 24–28 of the Act. Previously, an owner who wished to develop in the near future would be unlikely to take the risk of letting premises in the meantime if there were likely to be difficulties in obtaining possession when needed.

The application of the Act is limited to those tenancies falling within the definition provided by section 23(1):

> Subject to the provisions of this Act, this Part of this Act applies to any tenancy where the property comprised in the tenancy is or includes premises which are occupied by the tenant and are so occupied for the purposes of a business carried on by him or for those and other purposes.

The protection has been extended by the 2003 Order to include occupation or the carrying on of a business by a company where the tenant has a controlling interest or, if the tenant is a company, by a person with a controlling interest in the company.

A business is defined as 'a trade, profession or employment and includes any activity carried on by a body of persons, whether corporate or unincorporated' (s.23 (2)).

Tenancies to which the Act does not apply

Certain types of lettings are excluded by the provisions of section 43, including agricultural holdings, farm business tenancies, mining leases and certain public uses.

Tenancies of licensed premises are included by the provisions of the Landlord and Tenant (Licensed Premises) Act 1990, although it should be noted that many public houses are managed or subject to a licence agreement, which will serve to keep the premises outside the provisions of the 1954 Act.

The words used in section 23 must be considered carefully. The protection is available to tenants (the Act does not extend to licensees) who occupy the premises for business purposes (a tenant who gives up occupation or ceases to carry on a business no longer has the protection of the Act even though remaining a tenant). If the premises are sublet, the subtenant would be eligible for protection (subject to satisfying the other provisions of the legislation).

There is a wide interpretation of the expression 'business'. Use as a tennis club has been held to be a business (*Addiscombe Garden Estates v. Crabbe (1958)*), as has the use of a store in conjunction with a retail shop nearby (*Hill Property and Investment Co. Ltd, v. Naraine Pharmacy Ltd (1979)*), and that of a residential flat from which the tenant ran an importing business (*Cheryl Investments v. Saldanha (1978)*). But a maisonette let to a medical practitioner who saw patients there on rare occasions did not qualify as a property occupied for the purpose of a business (*Royal Life Saving Society v. Page (1978)*).

Agreement to exclude protection of sections 24–28

If both parties agree to enter into a tenancy on business premises for a fixed pre-determined term of years (a term certain) and to exclude the provisions of sections 24–28, there is a procedure provided by the 2003 Order for the parties to signal their agreement. Previously the parties were required to obtain a court order authorizing the exclusion before completion of the lease.

The implications of a 'contracted out' lease are severe in that at the end of a term the tenant will be unable to claim a new lease under the Act; even where the landlord is willing to offer a new lease, the tenant will be unable to invoke the powers of the court to adjudicate on those terms in the event that the parties are unable to reach agreement; no compensation will be payable for loss of occupation rights.

The rental value of the premises without the security of tenure afforded by the Act will tend to be lower than that for similar premises with the benefit of the Act; a tendency that may be reflected if an arbitrator or other conciliator is called upon to fix a rent on review. It has important implications for valuations of business premises.

As stated, the parties do not opt out of the whole of the Act but only sections 24–28. These relate to:

i   Continuation of tenancies and grant of new tenancies (s.24).
ii   Rent while tenancy continues by virtue of section 24. Any application for the determination of an interim rent under section 24A must be made within six months of termination of the relevant tenancy.
iii Termination of tenancy by the landlord (s.25); in the form prescribed by the 2003 Order.
iv   Tenant's request for a new tenancy (s.26), again using the forms prescribed by the 2003 Order.
v   Termination by tenant of tenancy for a fixed term (s.27). Section 24 will not be applicable in relation to a term of years certain where the tenant is not in occupation at the point when the tenancy would have come to an end by the effluxion of time.
vi   Renewal of tenancies by agreement (s.28).

Various notices are prescribed by the Order. Apart from requiring certain information to be given, they emphasize the need for the tenant to take professional advice.

Even when a tenancy is for a term certain, it continues as a periodic tenancy until determined at its expiration or later by one or other of the parties to the lease.

Where the tenant wishes to leave, notice should be given in accordance with the terms of the lease (but in any event not less than three months' notice to expire either at the end of the term or on any subsequent quarter day as provided in s.27).

Contracting out

Following the 2003 Order, the parties may contract out of the statutory process of agreeing a new tenancy. For the effective contracting out of a new tenancy the landlord must give a warning notice to the tenant in the prescribed form not less than 14 days before the tenant enters into the contract of tenancy, to which the tenant must reply with a declaration accepting the contents of the notice to the effect that sections 24–28 of the Landlord and Tenant Act 1954 do not apply to the tenancy.

There is a further provision to the effect that an existing tenant can agree to surrender the statutory protection following a similar procedure.

The consequence of contracting out is not only that the tenant has no right to claim a new tenancy on the expiration of the old, but that no compensation is payable on vacation of the premises at the end of the term.

There is an anomalous situation where a tenant of premises to which the Act does not apply can create a subtenancy to which the Act does apply, even if the subtenancy is unlawful. By contrast, an assignment of the tenant's interest would not provide such protection.

Tenant's application for a new tenancy

Should the tenant wish to stay but be anxious for the grant of a new term of years, an application should be made as provided in section 26, serving notice in the form prescribed by the Regulatory Reform Order 2003.

Where the landlord decides to serve a notice, either to obtain possession or simply to renegotiate the terms of the lease, the notice must be in the form prescribed in the regulations as provided by section 25. The notice should indicate whether the landlord is willing to grant a new lease; if not, the reason or reasons for not doing so should be specified and, to be valid, must fall within the provisions of section 30(1), subsections (a)–(g), which may be summarized as follows:

a   Breach by the tenant of the repairing covenant.
b   Persistent delay in paying rent due.
c   Other substantial breaches by the tenant of his obligations.
d   Alternative accommodation provided or secured for the tenant by the landlord (it must be suitable, reasonable and available at a time to suit the tenant's requirements including the requirement to preserve goodwill).
e   Uneconomic letting: where the current tenancy was created by a subletting of part only of the property and where the aggregate of rents on separate lettings produces a rent substantially less than the rent reasonably obtainable on a letting of that property as a whole.
f   Demolition or reconstruction of the premises or a substantial part of them. But by section 31A (introduced by the Law of Property Act 1969) the tenant agrees in the terms of the new tenancy to the landlord having access to carry out the work and that it could be carried out without obtaining possession or

that the tenant is prepared to accept a tenancy of an economically separable part of the holding.

g   Occupation by the landlord when the current tenancy comes to an end, either wholly or partly for a business carried on by the landlord or to be used as the landlord's residence. This ground applies only where the landlord's interest was purchased or created five years or more before the termination of the current tenancy.

Grounds (d), (f) and (g) are mandatory, while the others are at the discretion of the court.

Various decisions of the courts have underlined that substantial evidence will be required of the conduct alleged in grounds (a)–(c) and which, if proved, will deprive the tenant of occupation and any compensation otherwise payable under the Act.

Where suitable alternative accommodation is offered (ground (d)), the tenant is not entitled to compensation, underlining the fact that any payment is intended to compensate for loss of goodwill (a word used in the Act but not defined). It follows that one of the tests of suitability in this case is whether the tenant can transfer the existing goodwill to the alternative premises offered.

The remaining grounds, (e)–(g), will, if proved, entitle the tenant to compensation based on the rateable value. The multiplier is one except where all or part of the premises have been occupied for business purposes and for the same business for the previous 14 years and on any change of occupier each succeeded to the predecessor's business, in which case the figure is twice the rateable value of the holding.

Either multiplier represents a substantial sum and the landlord will wish to consider the implications. For example, in ground (f) where it is proposed to carry out reconstruction work where the tenant is willing to stay in the premises while the work is being done in accordance with provisions in section 31A(1), the landlord would then be absolved from paying compensation and so may wish to consider whether it is possible for the work to be done in this way.

Landlord's right to possession

Finally, where the landlord requires possession (whether to occupy and use as business premises or for these and other purposes), there must have been prior ownership for a minimum period of five years (s.30(2A)).

To be effective, notice under section 25 of the Act must state whether the landlord is opposed to a new tenancy and if so on what grounds or, if not opposed, to state the terms of a new tenancy. Where the tenant requires a new tenancy, an application must be made to the court before the expiry date given in the landlord's notice. The time cannot be extended.

Interim rent changes

Under the provisions of the 1954 Act, tenants benefited greatly by delaying settlement of the new rent because it was not backdated on settlement to the end of the previous tenancy (although any increase in market rent during that time would be reflected in the rent fixed). Section 24A was added to the 1954 Act by the 1969 Act to enable the landlord, by notice, to claim an interim rent. The effect of this was to give the opportunity for a revised rent from the beginning of the notice, reducing the benefit to the tenant.

Under the 2003 Order, either party may apply but must do so before the end of the continuation tenancy. The interim rent will now become payable as from the earliest date that could have been specified in either the landlord's or tenant's notice for a new tenancy to begin. The interim rent will be fixed at the same amount as the rent to be paid under the new tenancy with the proviso that either party may apply for a different interim rent to be fixed in three cases:

- a significant change in the market;
- a change in the letting terms where the interim rent should reflect the terms of the earlier letting;
- a combination of the above circumstances.

The interim rent is based on a rent from year to year having regard to the existing tenancy, which tends to produce a rent below market value by perhaps 10–20 per cent for a lease period, depending on decisions reached by the courts. However, the interim rent so determined is payable whether or not the tenant eventually proceeds with a new tenancy, so the tenant takes some risk in staying on without knowing the level of rent to be fixed.

The parties negotiate for a new tenancy where they both wish to continue or where the landlord's reasons for terminating are not upheld by the court.

Terms of new tenancy

Three sections of the Act relate to the terms and conditions of the new tenancy. Section 33 (as amended by the 2003 Order) provides for a maximum duration of 15 years and the court is empowered to include provisions for earlier review. However, while this limitation binds the court, a partial agreement by the parties for a longer term could be incorporated into the court order should they so agree.

Section 34 sets out the 'disregards' in relation to the market rent to be fixed, while section 35 relates to other terms of the tenancy.

Where it later appears to the court that in making an order for termination of the current tenancy without an order for the grant of a new tenancy or in refusing an order for the grant of a new tenancy that the order was obtained by misrepresentation or the concealment of material facts, the court may order the payment of compensation by the landlord. The basis of the compensation is to be such sum as appears sufficient compensation for damage or loss sustained by the tenant as the result of the order or refusal less any statutory compensation previously paid on termination.

Referral to court

Either party may have recourse to the court if agreement cannot be reached, but it is important to note that where the parties have reached partial agreement they may request the court to complete the agreement by ruling on the outstanding items only, incorporating the previously agreed items in the order of the court for a new tenancy.

Referral to third party

The Law Society and the Royal Institution of Chartered Surveyors (RICS) have set up a scheme to settle terms where a claim to a new tenancy is unopposed. It is referred to by the acronym PACT (Professional Arbitration on Court Terms). Where the parties consent, the court will stay the claim while the parties make a joint application to use the scheme. The scheme provides for the appointment of a surveyor or solicitor to act either as arbitrator or independent expert as the parties decide. It is claimed to be more flexible and speedier than court procedures. The person appointed is able to consider the amount of the new rent, the interim rent, the duration and other terms of the lease. The outcome is binding on both parties.

Compensation for tenant's improvements

Where the tenancy does not continue, any improvements carried out by the tenant under the procedure laid down by the 1927 Act will attract compensation. To be eligible, the landlord's consent to the work, or, failing that, a decision of the court, must have been obtained before the work was begun.

Where the tenant carried out the improvements, compensation on termination of the tenancy will be based on values and costs current at the time of the claim. The compensation is the lesser of the net addition to the value of the property as a whole or, if lower, the reasonable cost of carrying out the improvement at the termination of the tenancy less any costs of repair.

It will be appreciated that the tenant is unlikely to receive any compensation where the site is to be cleared and redeveloped, a consideration that may guide the landlord when deciding at an earlier stage whether to elect to offer to provide the improvements proposed by the tenant.

RESIDENTIAL TENANCIES

Residential property has a long history of statutory interference with the contractual relationship between the parties. Until 1989, most tenancies were subject to some form and level of rent control. Other parts of the legislation were directed towards giving security to tenants and rights of succession to certain members of the family.

Many landlords were faced with significant expenditure on repairs and maintenance while not receiving an adequate rental income to cover costs. As a result, most dwellings were sold once they became vacant.

The serious shortage of private rented accommodation brought about by this intervention was alleviated to some extent by the public provision of council housing between the two world wars and for some time thereafter. The accommodation was of a better standard than most of the tenanted accommodation in the private sector, particularly in that it provided bathrooms, indoor sanitation and a garden. The waiting lists were long, however, and not everyone qualified for acceptance on to the lists. There were no incentives for private investors to provide units and with the contraction of council house building and the subsequent 'right to buy' provisions, the total stock declined.

Attempts to make more accommodation available in the private sector by rewarding the landlord with a 'fair rent' defined to exclude scarcity were largely unsuccessful.

The situation was finally recognized with the enactment of the Housing Act 1988, which provided for the landlord to charge and maintain a market rent in respect of tenancies granted after 14 January 1989.

Some tenancies already in existence in 1989 continue as rent act tenancies, where the rent is restricted to a 'fair rent'. With the passing of time, most tenancies are now assured tenancies, where the landlord may charge a market rent and recover possession should the tenant fail to pay rent for more than two months.

The shorthold form of assured tenancy enables the landlord to gain possession on expiry of a fixed term. Where no notice is served to expire at the end of the fixed term, the tenancy will continue on a periodic basis.

Service charges

The Landlord and Tenant Act 1985 defined a service charge as an amount payable by a tenant of a dwelling as part of or in addition to rent and which is payable, directly or indirectly, for services, repairs, maintenance or insurance or the landlord's costs of management and the whole or part of which varies or may vary according to the relevant costs.

Improvements are not included in the definition except where expressly referred to in the service charge provisions of the lease. Some improvement is, of course, inherent in most repairs.

From 1 October 2007 any service charge demand must include a summary of the tenant's rights and obligations in a form set out by the regulations. The charge is not legally payable unless the demand complies with these provisions; interest charges or other penalties for non-payment cannot be imposed where the regulations have not been followed.

Service charge payments are required to be held on trust to the exclusion of any other provisions contained in the lease. Service charges payable to a local authority, New Town Corporation or the Development Board for Rural Wales are not affected by the provisions, unless the lease is for a period of 21 years or more.

Deposit protection scheme

Most landlords require a deposit from the tenant against any damage or rent out-standing at the end of the tenancy. Some tenants have experienced difficulty in obtaining a refund on termination. Since 6 April 2007 imposed landlords or their agents have a responsibility to protect any deposit paid by tenants as a condition of the tenancy.

Any deposit must be transferred to one of three groups authorized to run the scheme. One group runs a custodial scheme, while the others are insurance based. There is no charge to the tenant.

The scheme applies to all assured shorthold tenancies. The landlord's right to regain possession at the end of the tenancy will be lost unless the deposit has been protected. A landlord or agent failing to comply may have to pay the tenant up to three times the amount of the deposit.

All three schemes offer the tenant help in resolving any dispute in negotiating how much of the deposit should be returned.

Investment in dwellings to let

The purchase of dwellings to let is now well established as an investment area.

Funding is available and mortgage interest chargeable against income as a business expense, but only where the mortgage does not include any element of repayment of capital.

Where lettings are to tenants in receipt of housing benefit, the rent level will be subject to scrutiny by an officer of the Rent Service.

Rents are not subject to VAT.

Gas appliances must be inspected on a regular basis and a certificate of inspection obtained.

Should the owner decide to dispose of the property it would normally be wise to gain possession before it is placed on the market so as to avoid any complications likely to delay completion.

THE PLANNING FRAMEWORK

Planning permission is required for any development unless specifically excluded by statute. Development is defined as:

> carrying out of building, engineering, mining or other operations in, on, over or under land or the making of any material change in the use of the buildings or other land.
>
> *Section 22 Town and Country Planning Act 1971*

For this purpose, uses are grouped into classes as set out in the Town and Country Planning (Use Classes) Order 1987 and subsequent amendments. The classes are summarized in Table 5.1.

Table 5.1 Use classes for planning purposes

Part A: high street retail and service outlets

A1	Shops	Shops, retail warehousing, hairdressers, travel and ticket agencies, post offices, pet shops, sandwich bars, showrooms, undertakers
A2	Financial and professional services	Banks, building societies, estate and employment agencies, professional and financial services and betting offices
A3	Restaurants and cafés	Restaurants, snack bars, cafés – sale of food and drink on premises
A4	Drinking establishments	Public houses, wine bars and other drinking establishments (but not night clubs: see *sui generis* below)
A5	Hot food takeaways	Sale of hot food for consumption off the premises

---

Part B: other offices and industrial uses

B1	Business	Offices, research and development, light industry suitable for a residential area
B2	General industrial	General industrial
B3–B7	Revoked	
B8	Storage or distribution	Including open air storage

Part C: residential uses

C1	Hotels	Hotels, boarding houses and guest houses where no significant element of care included
C2	Residential institutions	Care homes, hospitals, nursing homes, boarding schools, residential college and training centres
C2A	Secure residential institutions	Including use as a prison, young offenders institution, detention centre, secure training centre, secure hospital, use as a military barracks, etc.
C3	Dwelling houses	Family houses or houses occupied by up to six residents living together as a single household, including a household where care is provided for residents

---

Part D: non-residential uses

| D1 | Non-residential institutions | Clinics, health centres, crèches, day nurseries and centres, schools, art galleries, museums, libraries, halls, places of worship, non-residential education and training centres |
| D2 | Assembly and leisure | Cinemas, music and concert halls, bingo and dance halls, swimming baths, skating rinks, gymnasia or sports arenas |

Table 5.1 continued

Part E: miscellaneous uses

Sui generis	Miscellaneous	Theatres, houses in multiple paying occupation, hostels (without any significant element of care), scrap yards, petrol filling stations and units selling or displaying motor vehicles, retail warehouse clubs, nightclubs, launderettes, amusement centres, casinos, taxi offices

Development takes place where the change of use is material, including the change of use from one use class to another. Change of use within a use class does not constitute development.

The following changes of use are excluded from the definition of development:

- Use of the buildings or land within the curtilage of a dwelling house for any purposes incidental to occupation.
- Use of any land for forestry or agricultural together with any buildings on the land.
- Use of land or buildings for any purpose specified in an order made by the Secretary of State.

There are detailed provisions for making applications for planning permission. Where an application for planning permission is refused or is granted subject to conditions, or where the planning authority fails to reach a decision within a specified time, the applicant may appeal. The appeal may be heard at a public local inquiry, but the parties may agree to proceed by way of written representations.

There are special provisions relating to advertising and caravan sites. Additional powers exist in relation to Conservation Areas, buildings of special architectural or historic interest (listed buildings), National Parks, Sites of Special Scientific Interest (SSSIs), scheduled monuments and areas of archaeological importance.

Planning enquiries should always be made prior to the formulation of a valuation since the information obtained may have a fundamental effect on value.

There are detailed regulations to enable the enforcement of planning control and any unauthorized development, breach of a condition of planning permission or use for an unauthorized purpose may result in service of an enforcement notice.

6   The comparative method

THE APPROACH TO COMPARISON

Comparison can be defined as 'the act of comparing' and to compare as 'to examine in order to observe resemblances or differences'. These definitions go a long way to explaining what the comparative method does. It works on the simple notion that if one property sells in the open market for £250 000 on today's date, then so should another, if it is exactly the same in every respect. But property is heterogeneous; no two are truly the same. Even if two neighbouring industrial units are built to the same floor area and specification, they will occupy different sites. In consequence, one may be further from the main road than the other and stand in its shadow in the sense that it lacks the visual prominence of its neighbour. A simple market-based comparison alone is insufficient. As the above definitions suggest, the valuer has to examine resemblances and differences and make appropriate adjustments to reflect these differences. In valuation, these differences can be described as the factors affecting value or the value determinants.

In a judgment by Forbes J, comparison was described thus:

> It is a fundamental aspect of valuation that it proceeds by analogy. The valuer isolates those characteristics of the object to be valued which in his view affect the value and then seeks another object of known, or ascertainable, value possessing some or all of those characteristics with which he may compare the object he is valuing. Where no directly comparable object exists the valuer must make allowances of one kind or another, interpolating and extrapolating from his given data. The less closely analogous the object chosen for comparison, the greater the allowances which have to be made and the greater the opportunity for error.
>
> *GREA Real Property Investments Limited v. Williams (1979)*

Capital value is determined by a number of factors: the present and prospective incomes; the return that the market determines to be appropriate to the particular investment; the strength of the tenant's covenant, the lease terms and the tenure of the property. It is therefore appropriate that in seeking to analyse real property transactions involving similar properties to the one to be valued, particular attention should be paid to these features. They are independent of each other but at the same time intricately interwoven; interpretation is often extremely difficult, involving aspects of judgement and experience.

ASPECTS OF VALUE

The factors affecting value can be split into legal, economic, physical and building-specific factors. The legal and economic issues were addressed in Chapter 5. The remaining ones are considered below.

Physical factors

Physical factors are concerned with the nature and extent of the property and its spatial context. Every building exists in space and will thus have an impact upon the environment in which it is placed and be impacted upon by that which surrounds it.

Location

Location is of paramount importance. The right property in the right location will always command a high value. Commuter homes near underground and mainline stations provide a residential example, while distribution warehouses near to major motorway junctions and interchanges are a commercial example.

However, location does not always dictate that central is the most valuable for any specific use. For example, a vegetable collection and packaging plant in the depths of the country may well be the most convenient place for such a building, despite its being a long way from a motorway. But should that use cease for some reason, it is likely to prove much more difficult to find a replacement tenant. Any change of use would probably present problems. It is unlikely that the unit could be recommended as a suitable investment.

The nature of use will affect choice of location. The distributor of high value portable electronic equipment may prefer the secure anonymity of an unmarked unit at the rear of a characterless industrial estate, whereas a prominent law firm insists on having a highly visible presence in a prestige location.

Topography

This concerns the lie of the land, the physical form of the land surface. A level site will normally be preferred. A steeply sloping site has its own difficulties in terms of construction, drainage and accessibility, including the needs of the disabled. Levelling a site can be an expensive business, especially if retaining boundary walls, and specialist civil engineering solutions are needed. Ongoing maintenance may well be more expensive.

Accessibility

This reflects the convenience of the site in relation to the existing road, rail, sea and air infrastructure. But it also has to do with the physical point or points of access to the site itself. Given the trend among many local authority highways departments to insist on separating car and heavy goods vehicle access to warehousing, a site with only one point of access may be at a disadvantage and thus less valuable than a site with two. Any physical constraints on the width or height of that access, such as a bridge, overhead power lines or proximity to a road junction, may all have an adverse impact upon value.

Geology

Geology differs from topography in that it considers the physical make-up of the ground. The natural geology may favour a site with a good load-bearing capacity for a variety of uses. However, fault lines, clays and sand may hamper development, requiring more expensive foundation design to overcome differential settlement and poor ground conditions. Nature alone is not to blame. Sites could be filled or made up, in some cases with contaminated material making them much less attractive for development and expensive to remedy. Features under the ground such as pipes, cables, culverts, tunnels and mine workings, such as shafts and galleries, can also create technical difficulties for the development and subsequent use of land to the detriment of its value.

Aspect

This relates to the positioning of a property within space. Thus a building in a hollow can be said to have a sheltered aspect, creating certain advantages over say an exposed coastal aspect where a property has to suffer the effects of strong winds, high tides, flooding and the relentless salt-water corrosion of ironware and steel fixings. This can be measured financially in terms of the increased costs of repairs, maintenance and insurance but also in terms of intangibles such as comfort and convenience in use.

Property specific factors

Property specific factors concern the nature and extent of a building and its general suitability for use, including construction and maintenance, age, condition, layout and specification.

Construction

This can have both positive and negative effects on value. Where durable and time-tested materials are used in conventional ways, a good solid building fit for its purpose results. But where new technologies are being used and the necessary skills needed for their fabrication are still at an experimental stage, it may be that future problems are being built into the property. Materials such as high alumina cement, asbestos, sea-dredged aggregates and woodwool boards used structurally are all collectively termed 'deleterious' in the sense of being harmful or injurious. The presence of such materials in a building would tend to have an adverse impact upon its value.

Age

The age of a property is interwoven with its construction in that the valuer cannot compare a Victorian terraced house with a new semi-detached house of modern specification. The value of the respective houses will reflect the age of the property in that, along with age, comes a greater need for repair and maintenance, usually more expensive for a larger property of Victorian proportions than for a brand new property where it is hoped no maintenance will be needed for at least the first few years. Likewise the brand new 'grade A' office with air conditioning, recessed lighting and information and communication technology (ICT) compatibility will be preferred to a 1960s refurbishment with boxed lighting, central heating radiators and perimeter trunking in which to locate computer cabling. This difference is not entirely due to the age of the property alone, but rather age intertwines with the specification, appearance and layout of a property in its impact upon value.

Condition

This will affect value in that the perceived costs of necessary repair and improvement will be reflected in any bid a purchaser will make to acquire the property and put it into a suitable condition, fit for the proposed purpose. Estimates of such costs will be deducted from the bid, which will also reflect an allowance for the delay and trouble of carrying out the necessary works. The valuer should exercise extreme care in attempting to compare the sale price of a semi-derelict shop with a recently refurbished one. There is a need to consider the condition of the property in terms of its physical condition, the decorative state of repair, and the suitability and safety of service connections.

Layout

The layout of the property is important. Long and thin is unlikely to be an efficient use of space except for a rope manufacturer. For most other uses, squares or rectangles represent the most economic floor plate from which to operate. Within that space there is little value in what can be termed 'dead space'. Typical of the 1960s was office space of a cellular nature with wide long corridors, spacious landings and stairwells; older office buildings do not make such efficient use of the space as their newbuild counterparts. Today's office occupiers require open spaces that can be partitioned to suit their own particular requirements.

The 'gross to net area ratio' is an efficiency measure. A typical 1960s building of 1000 m² gross may lose up to 35 per cent of its total area to corridors and other circulating space to the point where the net (useable) area extends to only 650 m². By contrast, a modern building of similar gross area but without the extravagant corridors, foyer, landings and stairs may lose only 15 per cent of the gross to net ratio, a difference of some 200 m² or one-fifth of the total gross area. The modern building thus has as much as 850 m² net upon which to base the value of the building.

Specification

This is an increasingly important factor in valuation, not just in office and retail sectors but also in industrial and warehousing. It can be defined as a detailed description of the construction, which has to do with the materials used and their suitability for the intended use. For example, a parcel delivery firm has a basic need for a warehouse. A shed with a surfaced yard may suffice. But how does their operation work and what vehicles do they use? If parcels are collected, brought to the building, sorted and then redirected, the building will have several specific requirements. It will need a cross-dock facility, i.e. a rectangular floor plan with numerous doors down both of the longest sides. Parcels arrive at one side of the building, are sorted in the middle and shipped out on the other side. In a large operation there may be a requirement for 15–20 doors along these longest frontages.

If the company operates large articulated vehicles, the floor level of the warehouse will need to be at the same height as the load bed of the trailer, allowing fork lifts and trolleys to be run from the building straight into the trailer. Given the variability in height of the load beds of heavy goods vehicles, mechanical dock levellers and electric or pneumatic height adjustable metal ramps will be required. Loading dock air curtains may be required to maintain a reasonable working temperature within the building.

To access such a technically advanced dock, an adjoining surfaced yard or 'apron' of a high specification is needed. Articulated vehicles require a large amount of space to manoeuvre in order to reverse. The sharp turning of the steering wheels and the dragging sideways of trailer tyres rips away at the surface such that tarmac is rarely a suitable surface. Much more durable is a fibre and steel reinforced power floated and grooved concrete yard. As this example illustrates, the valuer needs more than a passing knowledge of the requirements of different end-users.

TENANT SELECTION

Tenant mix

Tenant mix provides an interesting challenge for a property manager as, from small parades to regional shopping centres, the mix of tenants is important.

Fashion clothes retailers will not want to be next door to fish and chip shops for the obvious reasons of smell and litter. However, two jewellers or two or more shoe retailers often like to be neighbours as they create comparison shopping and complement each other, drawing a bigger crowd than would one such shop alone. The nature of the tenants and how well they work together is an extremely important factor for the individual tenant and for the centre.

Anchor tenants

Anchor tenants play a crucial role in the success of a new retail area. The fact that a well-known retailer has committed to space will encourage other traders to take space. Fosse Park on the outskirts of Leicester, just off junction 19 of the M1, is an out-of-town retail park where Marks & Spencer provides the main attraction for shoppers. This then provides the catalyst to draw in and retain other retail tenants who trade well as a result of the anchor tenant's presence. Sometimes it is the name alone that confers anchor tenant status; on other occasions it is the sheer size of their occupation. The Shires Shopping Centre in Leicester city centre has the department store Debenhams as its multilevel anchor tenant while other occupiers have single units on one or two floors. The significance of an individual tenant to the scheme as a whole cannot be underestimated.

ANALYSIS OF TRANSACTIONS

The dictum is 'analyse as you value, value as you analyse'. In other words, valuation as a process needs to be capable of being used as a tool of analysis as well as a tool of valuation. Comparison, then, not only values, but also devalues.

Units of comparability

Rent can be expressed in pounds (sterling) per square metre per annum. Purchase prices can be expressed in pounds (sterling) per square metre capital value. Investment sales can be analysed to reflect the return on the price paid to give a percentage yield. Here are three different units of comparison. Each is objective in that it reflects all of the factors affecting value.

MARKET EVIDENCE

As valuation can be defined as the estimation of price, it follows that the best market evidence is that of actual transactions. The valuer will be particularly interested in actual sales, lettings and investment transactions. As the market is not a centralized and open market such as the stock market, information is not always readily available in one place. But it is an important part of a valuer's activity to find information on relevant transactions and to analyse their components.

There is a variety of sources for such data. Agents' marketing boards with a SOLD slip across them are current and provide contact details. The property press, local and national, will report and perhaps even feature new deals. The specialist journals such as the *Estates Gazette* and *Property Week* include details of sales, lettings and investment transactions. Auction houses such as Allsops are a very useful source, as pre-auction catalogues generally provide a good deal of information. The Internet provides a wealth of valuation databases, with varying levels of reliability. The electronic versions of the professional journals provide searchable marketplace facilities to find and track property transactions. More general market information on rental and yield trends can be found on the websites of organizations such as the Investment Property Databank (IPD) and, in the case of residential property, of the main building societies. There are subscription services for research and market information.

The analytical approach adopted by the valuer will vary according to the type of property under consideration. The approach to residential property is often much more a function of local market knowledge and experience, whereas commercial property requires the detailed investigation referred to earlier.

APPLICATION TO MARKET SECTORS

Residential

Cited as the classic example of the application of the comparable method of valuation, residential property is typically a sector in which there is a wealth of transactional evidence upon which to fall back. 'For Sale' boards are a good indicator of market activity and a subscription to the Land Registry will allow the valuer to obtain reliable house purchase price data. However, the science of valuation blends with the intangibles of preference and sentiment. Good house valuers know their local market well. They will know which estates, even streets, have a good reputation and are therefore much sought after. They will know what features of the property the market will sustain. For example, in a certain locality a three-bedroom house extended to make four bedrooms will see an increase in value whereas a four-bedroom house extended to make five may have less of an increase in value in percentage terms.

It is essential to back up the intuitive feel for residential values with a proper analysis of the comparable evidence available as demonstrated in the following example.

Example 6.1

Following instructions to value number 30 Vale View Road, you have made an inspection of the house and neighbourhood and extracted office records showing recent sales. All the houses are modern and detached; number 30 contains three bedrooms, two reception rooms and has a double garage. A schedule of transactions separates odd and even numbers as the inspection confirmed that the even numbered houses enjoy unbroken views of the vale the street is named after. The odd numbered properties back on to a similar parallel street of houses. The office records are reproduced in Table 6.1.

When comparing the two three bedroom, two reception, single garage properties (numbers 11 and 14), it appears that one side of the road commands £10 000 or some 5.5 per cent more than the other.

Number 30 has three bedrooms, two reception rooms and a double garage. Thus the best comparable in terms of features is number 17, but it is an unhelpful comparable as it is not on the same side of the road.

Number 18 is similar but for the single garage. It would be useful if the valuer could isolate the contribution the extra garage space would add to the price, but the permutations of the evidence do not make this easy.

Numbers 14 and 18 are similar but for an extra bedroom and the difference in price is £30 000, suggesting that an extra bedroom adds £30 000 to the value. Applied to number 26, the £196 000 achieved should in theory increase to £226 000, suggesting that the value of number 30 is a similar £226 000. Number 17, on the opposite side of the road, has sold for £214 000, setting the minimum price for number 30. However, if the increment for views identified above is sustainable, then the £214 000 achieved for number 17 plus the increment of 5.5 per cent is £225 770. This serves to confirm a value in the order of £226 000.

The above example is of course totally contrived. In reality, any valuer would have to make further adjustments for age (of the property and of any comparables), condition, state and extent of the gardens and so forth. But the example does serve to illustrate a number of salient points.

The valuer should always seek to rely on more than one piece of evidence. In this example, the valuer used and adapted numbers 26 and 17 as comparables. The evidence for all the properties becomes a useful resource. The valuer would do well then to record in a database all these transactions together with the observation that they support units of comparison that suggest £30 000 per bedroom and, using the now valued number 30, a £6000 premium for an extra garage space (comparing numbers 30 and 18). Dates should be included with

Table 6.1 Office records of recent sales after sorting

Vale View Road

House number	Number of beds	Number of reception rooms	Garage: single or double	Purchase price (£)
11	3	2	S	180 000
17	3	2	D	214 000
14	3	2	S	190 000
26	3	2	D	196 000
18	4	2	S	220 000
32	4	2	D	250 000

the evidence to allow the valuer to consider trends over time, to ascertain whether values are rising or falling and to what extent.

Offices

The term covers a wide range of provision. Offices are united in that the means of measurement is net internal area (NIA). However, they vary greatly in nature, extending from the small provincial office above the local convenience store to the iconic multistorey landmark trophy building.

The quality of office accommodation will be of key concern to the valuer. Offices are often ranked by grade and will reflect their status as newbuild, modern (recently built), refurbished or old. They may occupy prime central business district locations, more secondary edge of town locations on out-of-town office sites, or be located on business parks. Each will have its own internal specification and the valuer will know what to expect for each grade of office in terms of facilities and specification.

The ground floor and lower floors will be expected to command higher values than the upper floors. However, a generous provision of passenger and goods lifts diminishes this differential; the extent to which it does so requires analysis of letting evidence. An office with its own car parking allocation will show higher rental values. It is likely that rental differentials will also exist for such things as air conditioning as opposed to central heating, raised access flooring as opposed to perimeter trunking, and low glare flush fitting lighting as opposed to suspended or boxed fluorescent tube lighting. While the well established price per square metre is the accepted comparable for offices, the detail behind it will reflect a thorough knowledge of market specifications and occupier requirements.

Retail

The purpose of a retail unit is to allow a trader the space and environment in which to sell goods. Increasingly, at one end of the market this will be the out-of-town shed or factory outlet unit. As these buildings are typically away from established town centres and within buildings more closely resembling industrial warehouses, rents and capital values are expressed in terms of pounds per square metre reflecting the overall NIA of the unit.

On the high street or in a shopping centre or precinct, the situation is very different. Here retailers are keen to secure a position with a good 'footfall', i.e. a high measure of pedestrian flow and a good, prominent glass shop frontage in which to display their wares and thus attract customers into the shop. This is especially so with clothes, shoes, jewellery and the like.

The frontage of the shop is important, as is the sales area immediately behind it, which is where a high proportion of trade is carried out. The quick purchase items are located here, while the specialist or bulky goods that require more time and thought to purchase are relegated to the rear part of the shop or even to upper floors. The reality of this theory can be confirmed in clothes shops: customer

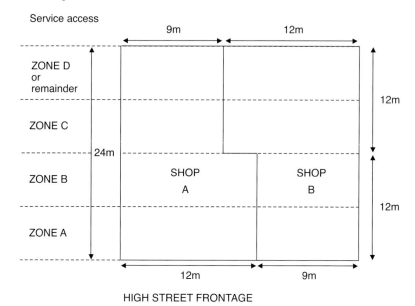

Figure 6.1 Layout of shops A and B

changing rooms and toilets would be unlikely to be located in the high turnover area at the front of the shop, but would usually be well down the shop depth, or at the rear, or on a different floor.

Over the years, the profession has developed an approach in which retail space is divided into imaginary zones for the purpose of analysing achieved rentals. The following example demonstrates an approach to the valuation of retail premises.

Example 6.2

Figure 6.1 shows shops A and B and their dimensions. Shop A has been let recently at a rent of £110 000 per annum. Table 6.2 shows the NIAs of the two shops to be exactly the same, but as one has a frontage that is 25 per cent less than the other, it is unlikely that both shops have the same rental value. It is important to discover the attitude of tenants to the disposition of space. Would a tenant prefer to have the same floor area but a greater frontage? One approach is to zone the retail area to reflect the relative value of the space according to where it is located within the unit. This form of analysis is termed zoning.

Table 6.2 Actual floor areas

Shop A		Shop B	
Dimensions	Area m²	Dimensions	Area m²
12 × 12	144	9 × 12	108
9 × 12	108	12 × 12	144
Total	252	**Total**	252

Zoning

Zoning is a means of analytical measurement that takes into account the configuration of space and the law of diminishing returns, enabling rents to be expressed in terms of Zone A (ITZA).

The zoning principle was developed when the typical shop unit was 20 ft wide and 60 ft in depth. It was usual to divide the depth into three zones of equal depth, although the practice varied in some parts of the country.

Following the move to metric measurement in 1996, most valuers now use a zone depth of 6 m, while the Valuation Office Agency uses zone depths of 6.1 m, the more accurate equivalent of 20 ft.

Typically three zones and a remainder, or four zones and a remainder, will be used, the valuer usually respecting any prevailing local market convention. So long as the valuer analyses and values in the same way, the choice makes no significant difference. But if, for example, rents are quoted on three 6.0 m zones and a remainder but applied to four 6.1 m zones and a remainder, an error will result. A diligent valuer will confirm the units being used. Due to the larger scale of central London retail, it is not uncommon to see larger zone depths used.

The zone immediately adjacent to the shop front is termed Zone A and represents the most attractive, and therefore the most expensive, retail space. Zone B is then assessed at half the value of Zone A. Zone C is the next zone, having a value half that of Zone B (and therefore one-quarter of the value of Zone A). Any additional depth is referred to as the remainder and is assessed at half the value of Zone C (i.e. one-eighth of the value of Zone A). For the purpose of analysis, the unit value 'halves back' through the zones. For example, a shop with a unit value of £1000/m² in Zone A would have a value of £500/m² in Zone B, £250/m² in Zone C and £125/m² in any remainder. Upper floors and basements used for retail purposes are often taken at a fraction of the Zone A figure.

To return to Example 6.2, a schedule (Table 6.3) sets out the information available. As stated previously, the two adjoining shops each have a net internal area of 252 m². Analysis using three zones and a remainder (at 6.0 m for ease of

Table 6.3 Analysis of areas and rental values

Zone	Shop A Area (m²)		Shop B Area (m²)	
	Actual	**ITZA**	**Actual**	**ITZA**
A	72	72	54	54
B	72	36	54	27
C	54	13.5	72	18
Remainder	54	6.75	72	9
Totals	252	128.25	252	108
Agreed rent	£110 000	857.70	Indicates rent of	£92 632
Overall rent/m²		£436.51	£367.59	

calculation) shows that each shop has the same NIA, but Shop A has an ITZA area of 128.25 m^2, while Shop B has an ITZA area of only 108 m^2. It must be emphasized that the ITZA area is not a physical area of measurement but a convenient one for the purpose.

Commentary

Referring to shop A, the rent is equal to an overall amount of £436.51/m^2, but in terms of Zone A the unit rent is £857.70. Shop B assessed on the basis of overall areas would also be worth a rent of £110 000. Applying the zoning principle, the rental value would be £92 632, an appreciable reduction but realistic given the narrower frontage and the perceived effect on trading. In practice, the valuer would wish to analyse several transactions before reaching any firm conclusions.

Adjustment would need to be considered where a shop had a return frontage to an adjoining street. That street could be a busy trading street, access to a car park serving the shopping centre or simply a service road. It is likely that the retailer occupying the corner site chose to take it because it offered a benefit. For example, shoe retailers and jewellers look for maximum window display space, while a baker or fishmonger might find the extra display area unnecessary or even counterproductive. The premium for the opportunity to have further window space requires judgement based on the circumstances, but is more difficult to back up with evidence of other transactions.

Natural zoning

The depth of zone is arbitrary but there is a broad consensus because zoning has proved useful as a way of comparing retail units over a fairly long period. Should the actual depths of shops in a particular location point to the benefit of a different allocation of depth, this should not present a problem (although the information would not assist in any attempt to establish a range of Zone A values over more than one centre).

Criticisms of zoning

Zoning provides a way of looking at space in an attempt to examine specific rental values. However, there are several criticisms of it, the main ones of which are listed below.

- It is an artificial process.
  Comment: it is an analytical tool of comparison for the valuer, not a decision-making tool for the occupier.
- It is irrelevant as retailers themselves do not use zoning to arrive at their rental bids but look at the prospects of making a profit, which they are able to estimate fairly accurately based on their experience of trading in other locations.

Comment: the valuer is not valuing the unit for a particular trade or occupier, but assessing a rent reflecting the size and arrangement of the floor space and the location of the unit within the centre.

- The zone depths are set arbitrarily by the valuer even though there may be no transactional evidence in that form to support a rent for a shop of any particular depth. This can be addressed by adopting what can be termed 'natural zoning'.

Comment: the approach is almost universally in use and therefore well tested. Its use does not prevent a ratepayer or tenant from producing alternative analyses and drawing conclusions relating to the immediate shopping area.

- It would be more appropriate to compare the relationship of net frontage to depth with rental value using regression analysis, which could be used to produce a 'line of best fit' relating the frontage to depth ratio and rental value.

Comment: the use of this statistical approach depends for its operation on the availability of additional data that is not always available.

Industrial

Collecting evidence for the valuation of industrial property is similar to that for other forms of investment. As with offices, the principle is one of comparing like with like, and in industrial property, rent and price per square metre remain the key units of comparison.

However, industrial property has changed from the days when 'industrial' meant 'factory'. There is a huge variety of industrial property, and care is therefore required in obtaining and interpreting evidence. The following example compares four units on the same estate.

Example 6.3

A recently developed industrial estate has found a ready demand. Information is available on four units, each of steel portal frames with steel cladding and with similar gross internal areas. The standard lease terms offer a 15-year lease with five-year upward only rent reviews on full repairing and insuring terms. The developer is prepared to sell any unit to show a return of 8 per cent. Information on available units and rents is set out in Table 6.4.

Table 6.4 Available industrial units: rents and prices

Unit	Area (m²)	Asking rent/ unit (£)	Asking rent/ m² (£)	Price to yield 8% (£)	Notes
A	465	22 000	47.31	275 000	
B	465	23 500	50.38	293 750	unit B includes
C	480	22 500	46.88	281 250	overhead
D	480	22 500	46.88	281 250	gantry crane

Table 6.5 Outcome of negotiations

Unit	Area (m²)	Rent achieved (£)	Rent/m² (£)	Sale price (£)
A	465	21 000	45.16	
B	465	–		
C	480	–		270 000
D	480	21 250	44.27	257 500

Commentary

There has been some progress in marketing as recorded in Table 6.5. Unit A has been let at £21 000 per annum (against an asking rent of £22 000). Unit B remains available. Accepting that the differential in rent of £1500 per annum between units A and B is a fair reflection of the rental value of the gantry crane provided in Unit B the rent likely to be achieved for the latter should be adjusted to £22 500.

Presumably the developers carried out some market research before deciding to provide the crane in Unit B; unless a tenant can now be found who needs the crane and is prepared to pay the additional rent, the decision to include it on a speculative basis may turn out not to have been very prudent.

Units C and D are similar in all respects but have different outcomes. The more complete information relates to Unit D where the developers secured a letting before selling the freehold to an investor. The fact that a tenant had entered into a lease enabled the investor to judge the quality of the covenant and to negotiate to achieve his target yield. It is known that the developers hoped to show a yield on any sales of 8 per cent but the investor has negotiated a a slightly lower purchase price to show a yield of 8.25 per cent. Acceptance of that offer by the developers suggests that the prospect of a sale as opposed to a letting was an attraction in assisting cash flow and reducing borrowings at that stage. The investor would have related his negotiating stance to what appeared to be achievable and to an assessment of the quality of the tenant's covenant.

With regard to unit C, the purchasers were searching for a suitable building to suit their requirements for business use.

Relationship between industrial and office space

There is a general market expectation about office space. In some markets, there is an expectation that an industrial property should have approximately 10 per cent office space. Where the office space is significantly more, market evidence would be sought.

Particular care is needed in the analysis and application of transactional evidence because of standard measuring practice.

The Royal Institution of Chartered Surveyors *Code of Measuring Practice* stipulates that industrial property is measured on a gross internal area basis to include office space where it could be considered to be ancillary to the industrial use. But, where the office space was located in a self-contained office building and used in connection with the industrial premises but capable of separate use, the proper basis would be to measure the offices on a net internal area basis. The relationship between office space and industrial use would require evidence of demand sufficient to affect the market either way. Similar considerations apply to the extraction of yield evidence from recent transactions.

Other property

Commercial property is typically categorized into office, retail and industrial sectors, although it includes a range of more unusual property such as hotels, petrol filling stations (with or without express shopping facilities), care and nursing homes and public houses. Although these specialized property types would normally be valued by reference to the profits method, any recent transactions of such types of property would be investigated in the quest for supporting evidence.

For example, a valuation of a nursing home would make reference to the potential fee income based on capacity and occupancy rates. Laing & Buisson's *Guide to the Healthcare Industry* provides reports of the weekly charges in every registered home across England by county. The same guide indicates the level of provision on a county basis.

It should be emphasized that the valuation of specialized properties is normally undertaken by specialist valuers who will have regard to income, costs and profits but will also look for comparisons with other transactions of which they have knowledge and which may enable a comparison to be made on the basis of price or rent per bed.

So even in the realms of quite specialist valuation, the comparable has a support role to play.

AUTOMATED VALUATION

There has been a strong movement towards streamlining the valuation process for the purpose of granting loans protected by a charge or mortgage on the borrower's home.

In certain circumstances lenders may accept:

- a 'drive by inspection' and valuation;
- a postcode valuation;
- an automated valuation process, adjusted to limit risk; the model relies on recorded comparables drawn from a variety of sources and restricts the amount borrowed.

With the gradual introduction of Home Information Packs (HIPs) recourse to these methods is likely to increase.

The Valuation Office Agency has investigated the grouping of property types for rating purposes, selecting a representative property and relating other properties in the immediate area to it. There is a safeguard in that any council tax payer can challenge the band in which their property is placed.

Mass valuation of the type described above is likely to grow, but relies to a large extent on the loan to value element. Any large-scale downward adjustment in the market would prove a testing time for these developments.

SUMMARY

- The comparable method involves the analytical use of market evidence and its application to the property to be valued.
- Property is heterogeneous and thus the valuer will have to account for differences between the subject property to be valued and the transactional evidence available.
- Comparison necessitates the use of units of comparison of which there are primarily three: direct comparison of capital values (such as in the residential market); comparison of unit values (either capital or rental unit values, mainly the latter); and market yields. In the case of retail space, zoning provides an invaluable tool of comparison.
- The valuer needs a good understanding of the factors affecting value, which can be grouped as physical factors (geographical and property-specific factors), legal factors (tenure and lease terms) and economic factors (supply and demand, time, risk and return).
- The valuer needs the analytical skills necessary to justify the adjustments applied to the comparable evidence.
- The comparable method is most suited to property types for which there is a wealth of market evidence. This comparable data should be captured and recorded to provide a database for valuation use in the short term but also for analysis of longer term market trends.
- As the comparable method is market-based, the valuer needs to know the market well, keeping up to date with occupier requirements, and must distinguish between owner-occupation and investment markets.

7 The valuation process

THE TRADITIONAL VALUATION APPROACH

When a valuation is commissioned, the valuer will wish to make a careful inspection and to discover information about the property to be valued that is not apparent from the inspection. Among the further property-specific details required will be information about the legal interest available, whether there are any leases, and if so their terms including rent, any planning or other proposals likely to affect the use of the property and the nature of the immediate neighbourhood. The valuer will be familiar with the market and the general influences being exerted on it and will hold or have access to details of recent transactions relating to sales and lettings of similar properties, which may be of assistance in determining the value of the property under consideration.

Valuers have a choice of approach when preparing a valuation. Some are committed to the traditional approach, while others advocate a discounted cashflow approach. The main distinguishing element between the two is that the traditional method values on the basis of current rents and values. Future rent increases are accounted for implicitly through the yield used to value the property. The discounted cashflow basis predicts growth and values on the basis of a market yield. The relative merits of the implicit and explicit approaches to valuation will be examined later.

THE ALL-RISKS YIELD APPROACH

A strong core of practising valuers continues to value investment properties using an implicit all-risks yield, notwithstanding the drawback that it is not possible to make direct comparisons with returns available in the wider investment market.

The all-risks yield may be defined as one that reflects implicitly all future benefits and disadvantages of the investment. Because the all-risks yield takes no explicit account of future rental growth, current rents and rental values are used in the valuations. There is no element of projection of rental values; any growth is inherent in the low yields adopted.

When valuing an investment property let on modern lease terms and at a full market rent recently agreed, there is direct market evidence of the all-risks yield (although it lacks any indication of the true anticipated return, accepting the low initial yield in the expectation that future rent increases will ensure a higher return).

Where the rent paid is historic and does not coincide with the current rental value, the approach attracts further criticism. The two blocks of income are valued separately; the rent for the remainder of the term is valued at a yield below the market evidence on the basis that it is more secure. The valuation of the reversionary element will take the full rental value into account from the time when it first becomes payable, but restricted to its level at the present time. That rental value is then treated as the income to be received indefinitely. This distortion of expected performance is again counterbalanced by using the all-risks yield, which is below current yields available on other types of investment, in anticipation of regular and substantial, but unquantified, growth.

It might be asked why such an apparently inappropriate yield is used. The response is that evidence of yields is gained principally from transactions involving rack-rented property, and the yield is found directly from the relationship between the net rent and the sale price. But each property is unique, even where superficially it appears to be the same as one on an adjoining site or in the next street. Consequently, the transactions selected as comparable and relevant will not be identical and intuitive adjustments will be required to reflect any actual or perceived differences.

The variations to be taken into account in any adjustment are the lease provisions including the repairing and insuring responsibilities, the rent review pattern, the age and general condition of the property, the suitability of the location, any special attributes of the building or its site, and the quality of the tenant's covenant.

Rack-rented and reversionary interests are considered within the traditional valuation approach. Reversionary interests include any arrangement where there is a difference between the current rent and the current rental value. In most cases, the current rent is less than the rental value because a review took place some time ago and another is not due yet.

There are some cases where the current rent exceeds the market value and others where the rent is fixed for a long period ahead, both of which warrant further investigation and a variation in the valuation approach to reflect what can be seen as limiting factors on value.

Leasehold interests present particular valuation problems and are dealt with later in a separate section.

FREEHOLD RACK-RENTED INVESTMENTS

The rack-rented investment property presents the most straightforward case. Where the rent is at full market value and the property is of freehold tenure let on modern lease terms, particularly with regular rent reviews at not more than, say, five-yearly intervals, the procedure involves capitalizing the rent in perpetuity at

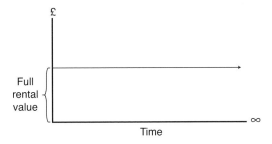

Figure 7.1 Valuation profile of rack-rented property

an appropriate yield derived from market observation and analysis. In other words, the right to receive a flow of income at the current level in perpetuity is capitalized using an all-risks yield derived directly from recent, relevant, market transactions. The assumed rent profile is depicted in Figure 7.1.

Example 7.1

Value a modern freehold office building occupied by the regional office of an insurance company where the property was let one month ago for a term of 25 years with five-yearly reviews on a full repairing and insuring (FRI) lease at a rent of £120 000 per annum. Recent transactions in the immediate locality suggest that investors are willing to accept an all-risks yield of 6 per cent.

Net rack-rent per annum	£ 120 000
Years' purchase in perpetuity @ 6%	16.67
Capital value	£2 000 000

Commentary

This is the simplest case. The rent is a recent one and is likely to represent the current rental value unless an initial premium has been paid. The rent being paid will be assumed to be receivable in perpetuity. The tenant is of undoubted status and management should be straightforward. Regular five-yearly reviews give an opportunity to rebase the rent to the then market level.

Details collected by the valuer include some information on market yields achieved in other recent sales, suggesting that an all-risks yield of 6 per cent would be obtained on a sale in the open market. Having made a thorough investigation, the valuer will have a high level of confidence in the valuation. Given the same information, other valuers would be likely to reach a similar conclusion.

The limitation of the valuation is that the yield is an all-risks yield with an implicit expectation that over its lifetime the investment will show a return in excess of 6 per cent. It is expected that the rental value will be greater at the time of the next review in five years' time and also at subsequent reviews. An investor would not purchase the property had the rent been fixed at its current level indefinitely or at least would not contemplate paying the current price for it. The investment is expected to show a better performance but the valuation does not quantify it.

Initial costs and annual expenses

It will be seen that no deductions have been made for the costs of purchase or for any recurring expenses.

The immediate response is that the comparables used to derive an appropriate yield were analysed in the same way; that is, initial costs and periodic management expenses were ignored. While this may be so, it is the case that an investor purchasing on the basis of the yield used will find that the investment has underperformed.

The initial transaction will involve legal costs, stamp duty land tax (SDLT) and valuation fees (together with VAT where applicable). The costs are not based on a fixed scale but will be influenced by the time taken and the complexity of the case. A reliable average figure is between 5.75 and 6.0 per cent of the purchase price. A large part of the expense is accounted for by SDLT, currently charged at 4 per cent on the total consideration where the purchase price exceeds £500 000.

Once acquired, most investors would expect to employ an agent to manage the property, specifically to collect rents, ensure the performance of covenants to maintain and repair, check that the insurance premium is paid up to date and that the building is covered for an adequate sum and for all necessary risks, negotiate rent on review (with the possible expense of taking the question to arbitration), monitor financial performance and advise on investment strategy. Taking one year with another, the average annual management costs in this example are taken to be 5 per cent of rent collected.

This further information is reflected in the following example.

Example 7.2

Using the information in Example 7.1 calculate the net return on the investment when initial costs and annual expenses are taken into account.

Net rack-rent per annum	£ 120 000
Deduct management @ 5%	£ 6 000
	£ 114 000
Years' purchase in perpetuity @ 6%	16.67
	£1 900 000
Deduct costs of purchase @ 6%	£ 107 547
Capital value	£1 792 453

Commentary

The effect of deducting management costs is of course a lower valuation. Had the investor paid £2 000 000, the value indicated in Example 7.1, the return would be slightly below 5.75 per cent. The one-off costs of purchase are likely to exceed £100 000: taking into account both sets of costs, the valuation suggests that the original figure is about 10 per cent greater than the figure now calculated.

The result is sufficiently different to suggest that such costs should be reflected. There has been no deception, but no transparency either. Where the costs are known or can be estimated with reasonable certainty, it is advisable to make deductions before analysing to find the yield.

It should be emphasized that costs should be reflected in this way only where the comparable evidence has been dissected and analysed making similar allowances. While analysing and valuing on a gross basis (i.e. ignoring annual management expenses and the costs of acquisition) is common and simplifies the calculations, the result gives an incomplete account of any transaction.

Note: in the valuation examples that follow adjustments for the costs of acquisition and management will be omitted. Each example is intended to demonstrate a particular point that will be clearer without the intrusion of information, which, while important, is not directly relevant to valuation principles.

FREEHOLD REVERSIONARY INVESTMENTS

The previous example was of a lease where the rent paid is at the full market value. Many investment valuations reflect a lease where the current rent is less than the full rental value because the last review happened a year or more ago. The investor anticipates an increase at the next review or at the end of the lease if there are no intervening reviews. The current rental value should be estimated using whatever assistance can be gained from information about recent lettings of similar properties. It is emphasized that no attempt is made to estimate the likely future rent. There are three recognized treatments of this type of valuation: term and reversion; layer (or hardcore); and equivalent yield.

Term and reversion (vertical separation)

The earliest treatment of reversionary investments was the term and reversion model. The yield is derived from analysis of market transactions, preferably of rack-rented investments. That yield is the basis for the valuation of the reversionary rent, which in most modern leases will make up the major part of the capital value. The term yield is adjusted downwards, typically by 0.5 or 1.0 per cent, to reflect both the tenant's contractual commitment to the current payment and the lower level of risk associated with it. This is a convention rather than a rule and is in the judgement of the valuer. It is not supported by market evidence, as there is no way of isolating that part of the capital value. There is some concern that, in

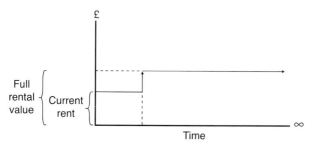

Figure 7.2 The term and reversion (vertical separation) profile

making the adjustment, the valuer is rarely fully aware of the impact it has on the overall return. The separation of income flows for valuation purposes is illustrated in Figure 7.2.

While it may seem a natural reaction to reduce a yield where the risk is less, there is no evidence that the market adopts such an approach. The risk rate is attributable to the property and not to a part of it. This approach also defies the claim that it mirrors the market, as there is no evidence that the market operates in this way. Where the market has indicated a preference for a particular yield, the valuer should reflect carefully before ignoring that preference and making an adjustment to one part of the income flow, even though in general the difference in capital value is insignificant. An example will demonstrate the process.

Example 7.3

A freehold shop property is currently let on an FRI lease with five-year rent reviews at a rent of £89 000 per annum with three years to run to the next review. The current rental value is £100 000 per annum; market evidence suggests a yield of 5.5 per cent.

Term rent per annum, net		£ 89 000	
Years' purchase 3 years @ 5%		2.7232	£ 242 365
Reversion to rack-rent, net		£100 000	
Years' purchase in perpetuity @ 5.5%	18.1818		
Present value of £1 in 3 years @ 5.5%	0.8516	15.4836	£1 548 364
Capital value			£1 790 728

Commentary

The valuer decides to reflect the certainty and the lower risk afforded by a rent below market value in the choice of a 5 per cent yield for the next three years. A new rent will be negotiated in three years' time. The figure for rental value is based on current values; it is the rent at which the shop could be let at the present time if not subject to the current lease. The expectation that the rental value will continue to increase is reflected in the low all-risks yield used.

However, the parties have done no more than agree to renegotiate at regular specified intervals, and while it is expected that when the rent is renegotiated in three years' time it will be at a figure higher than that used in the reversionary part of the above valuation, there is not the same quality of certainty about future events.

Example 7.4

Compare the capital value of the property described in Example 7.3 with the value of the same property using the same market-derived yield for both parts of the valuation.

Term rent per annum, net		£ 89 000	
Years' purchase 3 years @ 5.5%		2.6979	£ 240 113
Reversion to rack-rent, net		£100 000	
Years' purchase in perpetuity @ 5.5%	18.1818		
Present value of £1 in 3 years @ 5.5%	0.8516	15.4836	£1 548 364
Capital value			£1 788 477

Commentary

The difference between the two values is a fraction of 1 per cent. Using the same market yield for both parts of the valuation is probably justified where modern lease terms make provision for regular updating of the rent; it establishes the property as a 5.5 per cent investment and is preferable because the valuer is unable to establish the basis for a lower yield beyond the claim that some reflection of the greater security is warranted. This would seem superfluous where the tenant is of undoubted standing.

An alternative layout is to calculate the value of the property as if it were let at the full market rent, and then to deduct the current shortfall in rent for the period to the next review as shown in Example 7.5. This procedure has some advantages. It draws attention to the potential of the investment by showing the maximum market value when let at the full market rental value in perpetuity. It deducts the shortfall for the next three years to find the lower capital value due to the current rental level. It should be slightly easier to visualize, leading to a clearer approach to valuation.

Example 7.5

Value the property described in Example 7.3 by first capitalizing the current rental value in perpetuity and then deducting the shortfall between it and the current rent for the next three years.

Rack-rental value	£100 000	
Years' purchase in perpetuity @ 5.5%	18.1818	£1 818 180
Less shortfall for next 3 years		
(£100 000 – £89 000)	£ 11 000	
Years' purchase 3 years @ 5.5%	2.6979	£ 29 677
		£1 788 503

Commentary

Apart from a slight difference due to rounding, the outcome is the same as that in Example 7.4: the format shows the capital value based on the rack-rent reduced by some £30 000 to account for the lower rent paid for the next three years.

Layer (or hardcore) model (horizontal separation)

The vertical model previously described reflects the way in which the income is anticipated, and takes account of the change to a higher income when it happens. The variation now described adopts an arbitrary position whereby the current (core) income is assumed to continue into perpetuity, the increase (the top slice or marginal rent) being treated as a separate stream in perpetuity, but commencing at some time in the future. As a result there is an artificial horizontal separation of core and marginal incomes (Figure 7.3).

This model was originally developed in the 1950s and described in detail at that time by White. It is clear that it was intended to deal with a particular tax situation. It has been suggested that it may also be used when valuing turnover rents and for over-rented investments. The model has gained a certain status despite warnings against its use, particularly in its shortened form, by Trott and others. As

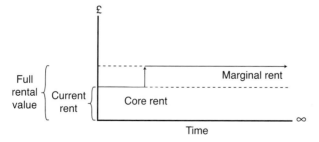

Figure 7.3 Layer of hardcore (horizontal separation) valuation profile

in the term and reversion model, the current rent and rental value are used: that is, there is no projection to show rental growth. The yields are calculated based on the market yield.

Example 7.6

Consider an office block let on FRI terms at £24 000 per annum with a reversion in three years' time. The current rental value is £40 000.

To find yield to be applied to marginal rent

Rack-rent		£40 000	
Years' purchase @ 6%		16.6667	£666 668
Current rent		£24 000	
Years' purchase @ 5.5%		18.1818	£436 363
Value of marginal rent			£230 305

$$\frac{(40\,000 - 24\,000)}{230\,305} \times 100 \text{ is } 6.95\%, \text{ say } 7\%$$

Hardcore or layer valuation

Core rent	£24 000	
Years' purchase in perpetuity @ 5.5%	18.1818	£436 363
Marginal rent	£16 000	
Years' purchase in perpetuity deferred 3 years @ 7%	11.6614	£186 582
total		£622 946

Compare with traditional valuation

Current rent	£24 000	
Years' purchase 3 years @ 5.5%	2.6979	£64 750
Reversion to	£40 000	
Years' purchase in perpetuity deferred 3 years @ 6%	13.9937	£559 748
		£624 498

Commentary

It is said that the layer method is much used in practice. It seems an unnecessarily complicated approach that does not easily relate to market evidence. The emphasis is on the current contracted income, which is carried into perpetuity. The increase is treated as less secure, and valued using a higher yield with the income deferred until the rent can be increased, in this case in three years' time.

The weakness of this approach is that the whole of the anticipated increase is valued at an appreciably higher rate, even though the rental figure used is the current one and below full market value.

It is claimed that the horizontal separation or layer method is more often used, even though this requires an arbitrary and unreal division of the income in the reversionary period, whereas the rent payable will be a single undivided sum. Where the conventional approach is to be used, the choice of the particular model should perhaps be made on the basis of which would be more readily presented to a client or even a tribunal or arbitrator.

The balance would appear to be much in favour of an equivalent yield usage to maintain a direct relationship with market information, and a vertically divided model, to ensure retention of a link with the expectations of rental change. As has been seen, investment valuations involving properties let on modern lease terms are unlikely to produce significant variations in the capital value.

Over-rented property

An over-rented property is one where the current rent is significantly higher than the current rental value. Such a situation is not common but, when present, raises relevant valuation concerns. Reasons will vary: possibly the tenant was badly advised when taking the lease; on the other hand, the rent agreed may have been a fair one when the lease was entered into but is now above the current rental value as it is in an area which, for whatever reason, is less popular than it was previously. There are many possible reasons why this should be so: for example, a new road bypassing a row of shops, the imposition of on-street parking restrictions, or the availability of superior premises nearby at competitive rents could all be reasons why a rental value has gone down.

The profile of an over-rented property is shown in Figure 7.4. The effect is that the income is less secure than is normally the case. It is possible that the deterioration will continue or that the market will improve until, at some time in the future, the rental value will equal and then exceed the current rent level.

The security of any rent in excess of the market value is difficult to judge, and depends to some extent on the quality of the tenant's covenant. An independent

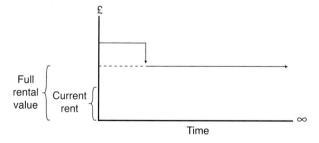

Figure 7.4 Profile of over-rented investment property

trader might well be unable to maintain the business if subject to higher overheads than competitors. Should the premises be vacated, the owner will have to accept the current rent level on a re-letting if a new tenant is to be found soon. The period during which the high rent is paid is a risky one, and any misgivings would be reflected in a valuation using a high yield to compensate for the disadvantages. Many investors would judge the property to be unsatisfactory as an investment and not consider it suitable to add to their portfolio.

Where there is a reversion or a rent review due within a reasonable time and the location is inherently sound, the problem may resolve itself by negotiation to a lower market rent. Where the tenant is of good standing and there is the possibility that the premises will be vacated, consideration should be given to offering to release the tenant from the current lease in favour of a new one at a lower rent, or negotiating a capital payment for accepting a surrender where feasible, although neither party is in a strong negotiating position. An over-rented property will appeal to a very limited investment market.

There is no model to suggest how to deal with a valuation but, as indicated, it is a situation that the landlord should seek to resolve if possible and that, if successful, is likely to prove to the advantage of both parties.

The next example sets out the traditional treatment; one of the contemporary approaches would probably be more relevant.

Example 7.7

A retail warehouse is let on a 30-year lease beginning in 1988 at a rent of £30 000 per annum on FRI terms. The site is an isolated one and since the building was erected, a retail park has been developed and the retail warehouse has been closed. The current value as a site to develop for another purpose is in the region of £60 000.

Prepare a valuation and advise your client, the freeholder.

Rent reserved	£30 000	
Years' purchase 11 years @ 12%	5.9377	£178 131
Reversion to capital value of site	£60 000	
Present value of £1 in 11 years @ 12%	0.2875	£ 17 249
Capital value		£195 380

Commentary

Using a market yield the property has a theoretical capital value of just under £200 000, although it is unlikely that a purchaser could be found at this level. It has little to commend it as an investment. The continued payment of rent relies entirely on the covenant of the tenant. The landlord would be well advised to open discussions with the tenant with a view to accepting a surrender of the lease subject to a negotiated payment. Redevelopment could then take place.

Equivalent yield

Any valuation of a reversionary interest by the traditional term and reversion approach uses two yields: one for the reversionary element derived from market analysis; and the other a lower rate to reflect the security of a rent below market level. The layer model in its original form also uses two yields, albeit derived in a different way.

There is growing support for use of the equivalent yield, which uses the same yield in valuing both tranches of the income flow. The reasons given are that analysis of any transaction can only give one yield that reflects market transactions: any attempt to vary the yields used introduces a subjective element for which there is no evidential support. The suggestion that a lower rent constitutes a lower risk that should be reflected in the yield is initially persuasive, but is difficult to sustain where both parts of the valuation are integral to the investment as a whole.

Where a valuation has been prepared on the basis of two yields, a calculation will reveal the equivalent yield.

The following example cites a sale based on a term and reversion valuation where there is a margin of 2 per cent between the two rates used. The equivalent yield can be found from tables, by trial and error, by formula or by a discounted cashflow calculation excluding growth.

Example 7.8

A factory is leased on FRI terms for a term of 15 years, of which 11 have now expired, at a rent of £11 500 per annum. The current market rental value on modern lease terms is £27 000 per annum. A sale has been agreed at £260 000 based on an all-risks yield of 9 per cent. Set out the valuation on which the sale price appears to have been based. Then calculate the equivalent yield using the various methods available.

The following information is required initially:

- the two yields used in the term and reversion valuation
- the value on reversion
- the gain on reversion.

Analysis of sale on term and reversion basis

Term rent per annum	£11 500	
Years' purchase 4 years @ 7%	4.1002	£ 47 152
Reversion to	£27 000	
Years' purchase in perpetuity deferred		
4 years @ 9%	7.8714	£212 528
Capital value		£259 680
(Sale completed at £260 000)		

The bulk of the value is in the reversion, therefore the equivalent yield will be nearer 9%

Initial yield $\qquad \dfrac{11\,500}{257\,630} \times 100 \qquad$ is 4.46%

and

Yield on reversion $\qquad \dfrac{27\,000}{257\,630} \times 100 \qquad$ is 10.48%

Check *Donaldson's Investment Tables*: approximate equivalent yield is 8.78%

Value by equivalent yield approach, using 8.75%

Term rent	£11 500	
Years' purchase 4 years @ 8.75%	3.2576	£ 37 462
Reversion to	£27 000	
YP perpetuity deferred 4 years @ 8.75%	8.1710	£220 617
		£258 079

Which is very close to the sale price.

Commentary

The analysis suggests that a 7 per cent yield was used to value the term income. The individual yields show a wide variation. The process of finding the equivalent yield is shown. The difference between the sale price and the reconstructed value on equivalent yield basis is below 1 per cent and the result sufficiently accurate for most purposes. The result could be found or checked by a discounted cashflow (without growth) analysis.

NON-CONFORMING PROPERTY INVESTMENTS

Rents fixed for long periods

There are some investments where the letting terms do not conform to those contained in the typical modern lease. It may be that there are no rent reviews or reviews only at extended intervals. Both situations require special consideration.

Investors are attracted to properties where the lease contains regular review provisions enabling the level of rent to be adjusted to the market value.

It follows that the absence in a lease of provisions to review the rent at regular intervals will have an effect on the capital value. Where the norm is considered to be five years, a period of, say, 10 years would be unattractive to potential investors. One of the reasons for investing in property is the prospect of significant growth; a lease for any length of time without a review would be seen as not conforming to that ideal. Most investors would expect compensation in the form of a yield premium before they would consider purchase, while many investors would be likely to decline the opportunity altogether.

It is argued that tenants should be prepared to pay an additional rent for a lease that either lacks a provision for review or where the reviews are at longer intervals than are usual. There is little evidence that tenants are prepared to act in this way; rent is part of their total outgoings and a higher rent would adversely affect their competitiveness.

When negotiating a lease, presumably the landlord obtains the best terms available from the potential tenants and the failure to negotiate regular reviews raises questions about the suitability of the property and the demand for it. Some early leases for long periods of 21, 35 or even 42 years fixed the rent for the whole period, the main concern being to ensure that a good quality tenant was found. It is known that, in addition, the tenant sometimes had the benefit of a provision that the lease could be renewed on the same terms except with regard to rent.

Where these or similar conditions exist, the rent will continue unchanged for a considerable period. Such an income is regarded as 'inflation-prone', reflecting more the features of the fixed-interest investment market than those usually associated with property and income growth. An investor in property would be reluctant to contemplate such an investment and would wish to know when it would be possible to regularize the position by entering into a new lease on modern terms.

The valuation should recognize the fixed non-growth term income by discounting at a market or non-property yield. The effect of this treatment is seen in the following example.

Example 7.9

A small office block is currently let on FRI terms at a rent of £30 000 per annum for the remainder of a lease expiring in 12 years' time. The current rental value is £60 000. Calculate the current capital value.

Term rent		£30 000	
Years' purchase 12 years @ 11%		6.4924	£194 772
Reversion to		£60 000	
Years' purchase in perpetuity @ 6%	16.6667		
Present value of £1, 12 years @ 11%	0.2858	4.7633	£285 800
			£480 572

Commentary

The rent is well secured, being fixed for the next 12 years. The current rental value of £60 000 per annum will have increased considerably by the end of the lease. The fixed nature of the term rent makes the initial part of the investment inflation-prone, recognized in the high yield used to value the term part of the investment. The relative unattractiveness of the investment may not be fully reflected in the valuation, even with the higher yield applied to the term value.

LEASEHOLD INTERESTS

So far, the valuation approach described has been limited to freehold investments. However, there are leasehold interests where the profit rent – the difference between what the leaseholder pays to the superior landlord and what is obtained from the subleaseholder – is of sufficient size and duration as to be a suitable form of investment.

Where the rent of a leasehold interest is less than the rental value, the difference between the two is referred to as a 'profit rent' in the hands of the leaseholder. The profit rent may be enjoyed for the whole of the remainder of the term, or only intermittently if there is no coincidence between the head and subleases; it may retain a constant relationship where such provision is made by the lease, or the financial relationship may vary primarily due to the lack of coincidence between head and subleases. Such complicated relationships increase the difficulty of analysis and valuation.

The terminable interest in leasehold properties raises problems of valuation not present in valuations of freehold interests.

By definition, all leasehold interests are of a finite duration. The length of the unexpired term is critical and will influence the valuation approach. Leases may be granted for very long periods such as 999 years, although most are for much shorter periods. For the purposes of valuation, leaseholds are referred to as short, medium and long. There is no precise definition, the main differentiation being between those interests with an unexpired term sufficient to ignore the limitation in valuation terms, and shorter terms where the interest is regarded as a wasting asset.

Where interests have 60 years or more to run, the practice has been to treat the income as received in perpetuity, as the capital value of any reversion after that time is likely to be a relatively insignificant part of the total capital value. Where the interest is for a shorter term, the convention is to assume the creation of a notional sinking fund to ensure replacement of the original capital. It is argued that it is then possible to derive the yield from similar, freehold investments where there is much more information. But the investment remains a leasehold investment and the freehold yield is therefore used only as a base, typically being increased by one or two points.

To offset the disadvantage of a wasting asset, a notional sinking fund is provided to replace the original purchase price at the end of the term, thus perpetuating the income and again seeking to justify the comparison with a freehold interest.

It is assumed that the amount reserved for the notional sinking fund will be taken from the rental income, and provision must be made for a sufficient sum to be available after tax. The yield on the sinking fund is taken at a low, safe rate to ensure that there is no risk of default, ensuring that the capital is replaced. The requirement not only for a safe investment but also for a guaranteed level of interest on sinking fund payments over a long term would attract relatively low rates. Net rates of 3–4 per cent are in common use. Investments with a short life would use the major part of the income in providing a sinking fund.

Although there is no evidence that sinking funds are taken out and used to replace capital, their provision in the valuation framework provides a theoretical base for comparison with freehold. The formula for years' purchase takes on a special form and calculates the years' purchase at the remunerative rate with provision for a sinking fund at the (lower) accumulative rate and adjusted for tax.

Where there are changes in the profit rent over the period of the investment, the valuation will proceed by stages, each with its own years' purchase multiplier and each providing a sinking fund. There is the prospect of two or more notional sinking funds based on replacement at the end of the respective periods; when one stage finishes and the next begins, the previous fund closes but is not called on because at that stage the investment continues to provide an income. The accumulation of interest on the closed fund ensures that there is a total overprovision of sinking funds. Some techniques of adjusting this overprovision are available. The better alternative appears to be a discounted cashflow investigation: however, there is little information about what rate of return is acceptable in return for a complete loss of the original investment. Much will depend on the size of the profit rent, the period for which it is to be enjoyed, and the attitude of the particular investor to risk.

Example 7.10

An office building is held on lease for an unexpired term of 15 years without review at a rent of £10 000 on FRI terms. The current rental value is £17 500 per annum. An appropriate freehold yield on a rack-rented property is 6 per cent. Interest rates on safe long-term investments are in the region of 4 per cent, and the appropriate tax rate is 25 per cent. Value the leasehold interest.

Leasehold interest	
Rack-rental value	£17 500
Less head rent	£10 000
Profit rent	£ 7 500
Years' purchase 15 years @ 8.5% and 4%, tax 25%	6.5968
Value of leasehold interest	£49 476

Ignoring dual rate and tax adjustment on sinking fund:

Profit rent	£ 7 500
Years' purchase 15 years @ × % single rate	6.5968
	£49 476

Which is equivalent to a return of approximately 12.5%

Commentary

There is a profit rent of £7500 per annum received for 15 years, after which the superior landlord resumes possession and the income to the present leaseholder ceases. It is assumed that an investor would require a yield of 8.5 per cent.

The leaseholder has a fixed rent for the remainder of the term. The rental value will continue to increase with an effect on the profit rent.

The problem is that the original purchase price is the amount replaced. The investor will be able to purchase another investment yielding roughly the same initial level of income. In other words, the purchasing power of the investment will not be maintained.

It is likely that a similar amount applied to the purchase of a freehold investment and not therefore needing to be replaced as a wasting asset would have a much higher value over the same period. An investor might well ask whether it is worth buying such a short-term investment and conclude that it depends on how good a return can be obtained. A further consideration is the situation at the end of the lease with a potential for a dilapidations claim.

The interest valued in Example 7.10 shows a good gross return (disregarding the provision of a sinking fund) of around 12.5 per cent per annum for 15 years, but, of course, no residual capital value.

In general, investors will find the decision between a limited-term investment and a permanent investment much simpler presented in this way: a higher annual return but the loss of capital. It will always be the case that had the original investment been made in a freehold interest, the freehold investment would outperform the leasehold one, although the initial income may be lower.

An alternative to the dual rate leasehold approach is to discount the anticipated flows of income at an appropriate equated yield, reflecting any rental growth expected where reviews exist to take advantage of increased rental values. The rent payable to the freeholder is then treated similarly and deducted from the earlier total.

Example 7.11

Modern shop premises are held on lease from the freeholder for an unexpired term of 13 years without review at a rent of £10 000 per annum. The premises are sublet for the whole of the term, the present rent being £12 500 per annum. The sublease is subject to five-year reviews, the next in three years' time. The current rental value is £14 000. Similar freehold investments change hands at yields in the region of 6 per cent. The equated yield may be taken as 11 per cent.

Leasehold interest	
Rent reserved	£12 500
Less head rent	£10 000
Profit rent	£ 2 500

Years' purchase 3 years 7.5%, 4%, tax 25%		1.9915	£4 979
Reversion to		£14 000	
Less head rent		£10 000	
		£4 000	
Years' purchase 10 years 7.5%, 4%, tax 25%	5.3748		
Present value of £1, 3 years @ 7.5%	0.80496	4.3265	£17 306
			£22 285

Commentary

The traditional approach to the solution of this more involved leasehold interest does not deal adequately with future increases in rental value, and a discounted cashflow calculation would directly reflect the future rent reviews.

Leasehold valuations are often undertaken on behalf of tenants wishing to assign their lease or surrender the lease in exchange for a lease for a longer term and possibly with changes to other terms. These negotiations often take place against a background of a transfer of the business and the sale of goodwill. It is likely that valuers will be appointed to act for each party. Following the changes to business tenancies introduced by the 2004 Order, valuations will vary according to whether the tenant has the protection of sections 24–28 of the Landlord and Tenant Act 1954 as amended. Without that protection, the value of any assignment would be likely to be seriously affected as the goodwill of the business would not be protected beyond the end of the current lease.

Relatively short leasehold interests may have a special appeal for non-taxpayers, for whom the benefit of the income is further enhanced.

Any investor in leaseholds, particularly shorter leaseholds, will appreciate the complications but may still be attracted to a greater return for a limited period. A short leasehold interest is an unusual form of investment and one that the investor needs to consider carefully. The decision will be more than usually interwoven with the investor's personal situation and wishes.

There is a possibility that some investors are attracted by the relative lack of competition and see the opportunity of appreciably higher returns. For example, a retired investor might see the benefits of a high return for a period of 5 or 10 years, accepting the eventual loss of any income.

A modest holding may sometimes be the key to a larger transaction, and the prospect of a third party being delayed by the intervening interest could result in an offer to buy out the interest out of scale with the real value of the holding. For example, take the case of the lease of a small workshop expiring in five years' time where the freeholder of the estate of which the workshop is a small part has had an offer to purchase the whole site for residential development. Where the workshop lease is the only thing standing between the freeholder and a substantial windfall profit, the leaseholder would be in a very comfortable position. It is not suggested that such a situation would present itself as part of acquiring a lease, but there is a possibility that the lease would have a value for the reasons suggested.

Many investors would not consider purchasing a short leasehold interest.

CONTEMPORARY ISSUES

Implied Rental Growth

Low initial all-risks yields are acceptable to investors in property only because there is an expectation of rental growth. Traditional valuations do not quantify the growth; they simply reflect what, on analysis of recent transactions, the market has accepted as appropriate, given past experience of market growth. That experience is likely to inform, at least in part, the level of expectation of future growth.

There will be a market yield for investments in general, adjusted for particular sectors and for individual investments within each sector. Once that market yield is known, the implications for growth can be exposed.

In some respects, investment in good quality modern property occupied by major commercial or industrial concerns is likely, in its own way, to be almost as default-free as is investment in government stock. The opportunity to renegotiate the investment return at regular intervals from a guaranteed base in the shape of an upward only review provision is unique to property investment and an extremely valuable attribute. Should the occupying tenant fail, there is a loss of income for a relatively short term until possession is obtained and another tenant found. In similar circumstances for shareholders in that same company, a catastrophic failure resulting in the liquidation of the company would leave them with no more than a share of whatever the liquidator was able to salvage from the remains of the company after paying all its other debts.

Implied income growth can be calculated given the all-risks and equated yields. On a yearly basis, the calculation is simple. Where the market yield is, say, 10 per cent and the return on the investment is 6 per cent, the income growth is the difference between the two, namely 4 per cent. But as investments in property do not usually afford an opportunity to make annual adjustments, rental growth must be related to the length of the review interval. The cumulative effect of the delay in revising the rent will be to require a higher rate of growth than if an annual adjustment could be made. The calculations necessary to identify and quantify growth will be used in the answer to Example 7.12, using the formula set out below.

$$K = E - (ASF \times P)$$
$$\text{where } P = (1+g)^{nt-1}$$

where

K = the all-risks yield (decimal)
E = the equated yield (decimal)

P = the rental growth over the interval
ASF = the sinking fund to replace £1 at 'E' for period 't'
g = annual growth (as a decimal)
t = rent review period.

Example 7.12

An investor has purchased a retail shop premises in the main thoroughfare of a town for £1 000 000. The unit has just been let to a national multiple on FRI lease terms for 20 years with reviews at intervals of five years, at a rent of £50 000 per annum. The market rate is generally agreed to be 10 per cent. Calculate the implied income growth.

Analysis of purchase	price	£1 000 000
	rent p.a.	£ 50 000
	years' purchase	20

Implied rental growth where market rate is 10% can be calculated as follows:

$$1 + g\char`^t = \frac{\text{YP perp @ K less YP 't' @ E}}{\text{YP perp @ K} \times \text{PV 't' @ E}} = \frac{20 - 3.7908}{20 \times 0.620913} = \frac{16.2090}{12.4184} \quad 1.305241$$

$1 + g\char`^5 =$	1.305241	Growth over five years (%)	30.5241
$1 + g =$	1.054722		
$g =$	0.05472	which is	5.47 % per annum

Commentary

The expectation is that the rental value of the shop premises will increase each year by 5.47 per cent, and that this will be reflected in substantial increases at each review. The annual growth is found from the formula as set out in the example.

It is sobering to reflect on the implication of the recent purchase, which is that the annual rental value will increase by an average of 5.5 per cent into the future.

DISCOUNTED CASHFLOW

Many investors and financial and academic commentators have expressed impatience and frustration with the continuing use of the implicit 'non-growth' approach to valuation. It is claimed that not only does it misrepresent the property market and make it difficult to know the effect of any yield adjustment, but also, and perhaps more importantly, it isolates property investment from other forms of investment in that the calculated yields from each are not directly comparable.

It does seem unsatisfactory to assume the continuation into perpetuity of rent at a level that would be unacceptable, capitalized at a yield that grossly understates the level of return expected. That the fictions may compensate each other to determine the 'correct' value is considered less than satisfactory by many professional investors who are keenly interested in the rate at which rents are expected to rise and in the current and projected return on capital invested.

The discounted cashflow (DCF) technique, a tool long used by accountants and business analysts, has been adapted for use in the appraisal of real property. The technique needed adaptation as it is now being required to deal with very long time scales, whereas the business use of the method is normally for a short 'pay-back' period only, which could be as little as 5 or 10 years, with the terminal value most often relatively low (scrap value or nil).

The two approaches within the DCF technique are the net present value (NPV) and the internal rate of return (IRR). NPV enables all the cashflows (incoming and outgoing) to be discounted at a selected rate of interest. When all the cashflows are summed (having regard to their signs) the result will show whether the target rate of interest will be achieved (0) or exceeded (a positive balance). A negative result shows that the particular discount rate specified will not be achieved.

Information is needed about the rent payable, the current rental value, the probable rental growth, the review intervals and the discounting rate of interest.

Valuers wedded to conventional approaches criticize the attempt to predict rental values and argue that the current rental value is as far as the valuer should go in providing information for a valuation. At the same time, they see no problem in using an all-risks yield derived from property transactions even though it bears no direct relationship to other non-property yields; they are content to adjust that yield up or down to compensate for perceived differences in age, quality and earning capacity despite their inability to gauge the precise effect of any adjustment on actual return.

One of the advantages of the DCF technique is the discipline of quantifying the anticipated rate of rental growth, which may cause the valuer to think more deeply about the qualities of the investment.

The selection of an appropriate yield should cause less difficulty since a direct comparison may be made with other investment media. The surrogate for yield comparison has been the gilt-edged market. Long-term or undated gilts were regarded as the perfect risk-free investments, any other form of investment being inferior and with a level of risk, the difference being reflected in an addition to the rate of return.

The definition of risk has widened considerably in recent years to include, in particular, the problem of maintaining the purchasing power of income. Gilts today are more properly described as 'default-free'. In the case of high quality investment properties let to blue-chip tenants on modern leases with upward only rent reviews, it could be argued that there is no default risk and that the overall risk is significantly less than that in most investments in equities.

However, there is risk in holding any long-term investments when compared with short-term Treasury bills or similar investments: risk for which an investor will expect to be compensated by way of an increased yield. However, real returns are likely to remain fairly low given the continuing level of inflation and the quantity of funds, including international funds, seeking investments of quality.

Whereas the NPV assesses an investment relative to a selected yield, the IRR determines the return produced by a particular income profit, reported as a yield, not an amount. The latter approach is therefore useful in non-market appraisals making use of the investor's required target rate or opportunity cost, as well as enabling an investor to compare two prospective investments.

Application of DCF techniques

The conventional valuation performed using an all-risks yield is a discounting exercise, summing the value of the right to receive successive incomes for a stated period. The convention is to use a low yield to compensate for the expectation that income will rise, linked to the use of the current rental value with no attempt to extrapolate the anticipated growth.

The discounted cashflow approach may be used in this way, but its distinguishing feature is that it is more often used to reflect capital value using a market yield (derived from the overall investment market but possibly adjusted to reflect the peculiarities of property) combined with income in which anticipated growth is quantified. The following calculations show the two approaches.

Commentary

Given the all-risks initial yield and the appropriate market yield, the implied rate of growth may be calculated.

Example 7.13

Your client owns the freehold interest in well-situated shop premises recently let on an institutional type lease at a rent of £25 000 per annum for a term of 25 years with five-yearly reviews. The all-risks yield is found to be 6 per cent from analysis of recent sales of comparable properties; the equated yield required is to be taken as 10 per cent.

Use trial rates of 9% and 11%

Year	Rent (£)	Amount £1 5 yrs@4%	Rent on review (£)	YP 5 years at 9%	PV £1 at 9%	Present value (£)
1 to 5	25 000			3.8897		97 243
6 to 10	25 000	1.2167	30 418	3.8897	0.6499	76 893
11 to 15	25 000	1.4802	37 005	3.8897	0.4224	60 800
16 to 20	25 000	1.8010	45 025	3.8897	0.2745	48 074

21 to 25	25 000	2.1911	54 778	3.8897	0.1784	38 011
26 to perp	25 000	2.666	66 650	16.6667	0.1160	128 857
						449 877
				Less price		416 668
						−33 209

Year	Rent (£)	Amt £1 5 yrs @ 4%	Rent on review (£)	YP 5 years at 11%	PV £1 at 11%	Present value (£)
1 to 5	25 000			3.7908		94 770
6 to 10	25 000	1.1593	28 983	3.7908	0.5935	65 206
11 to 15	25 000	1.3439	33 598	3.7908	0.3855	49 098
15 to 20	25 000	1.5580	38 950	3.7908	0.2394	35 348
21 to 25	25 000	1.8061	45 153	3.7908	0.1486	25 435
26 to perp	25 000	2.0938	52 345	16.6667	0.0923	80 524
						350 381
				Less price		416 668
						66 287

$$\text{Equated yield} \quad 9 + \left[\left[\frac{33209}{33209 + 66287} \right] \times 2\,(11\% - 9\%) \right] = 9.67\%$$

The valuation is straightforward: the shop is rack-rented and there is market evidence of the all-risks yield.

The capitalized amount is deferred at the equated yield rate. Where rental growth is known the equated yield can be found by taking two trial rates. The values are then calculated for both rates, one of which should give a positive result and the other a negative one. In that event, the true yield is captured between the two yields and may be determined more precisely by formula or by selecting two more yields and repeating the calculation until the 'correct' yield is found when the sum of the discounted incomes is equal to '0'.

Example 7.14

A property with a current rental value of £20 000 per annum and let on FRI terms has been sold to show an all-risks yield of 6 per cent. Calculate the internal rate of return.

Rental value	£ 20 000
YP perp @ 6%	16.6667
Capital value	£333 333

Period (years)	Rent (£)	Growth at 4 %	YP 5 years at 8.5 %	Value (£)	PV £1 at 8.5 %	NPV (£)
1 to 5	20 000		3.9406	78 812		78 812
6 to 10	24 334	1.2167	3.9406	95 891	0.6650	63 767
11 to 15	29 604	1.4802	3.9406	116 658	0.4423	51 598
16 to 20	36 018	1.8009	3.9406	141 933	0.2941	41 742
21 - perp	36 018		16.6667	600 301	0.1956	117 419
				Total		353 338
				Deduct value		335 000
						18 338

Period (years)	Rent (£)	Growth at 4 %	YP 5 years at 9.5%	Value (£)	PV £1 at 9.5%	NPV (£)
1 to 5	20 000		3.8397	76 794		76 794
6 to 10	24 334	1.2167	3.8397	93 435	0.6352	59 350
11 to 15	29 604	1.4802	3.8397	113 670	0.4035	45 866
16 to 20	36 018	1.8009	3.8397	138 298	0.2563	35 446
21 - perp	36 018		16.6667	600 301	0.1628	97 729
				Total		315 185
				Deduct value		335 000
						−19 815

$$\text{Find IRR} \quad 8.5\% + \left[\frac{18\,338}{18\,338 + 19\,815} \right] = 8.98\% \text{ (say 9\%)}$$

Commentary

As stated, the sale shows a return on an all-risks yield basis of 6 per cent. When the expectation of growth is taken into account, it is clear that an overall yield of 9 per cent would be achieved if the forecast rental growth was confirmed by events.

Real value/equated yield hybrid

With growing inflation, the shortcomings of valuations using all-risks yields raised some concerns. Several criticisms from outside the profession gained a good deal of publicity and support and generated a wider debate.

There was a wider sensitivity because it was around the time of the secondary banking crisis that involved the Bank of England in financial support for institutions and others overstretched by lending on, among other things, property investments.

Academics were aware of the shortcomings and had already been considering the problem, but without much progress. Ernest Wood suggested a 'real value model' that he promoted with some energy, but it was not taken up. However, the general concern had touched a nerve among the leaders of the profession and the Royal Institution of Chartered Surveyors (RICS) appointed a research fellow,

Andrew Trott, to investigate the methods of valuation used in practice, resulting in two reports, neither of which was much noticed outside the academic group of valuers. The reports were helpful in reviewing the current methods, but little progress was made in resolving the problems caused by high levels of inflation, although greater use of discounted cashflow methods were encouraged. Trott's interim report suggested that Wood's 'real value' model was too complex for most valuers to be able to use in their day-to-day work. This seemed more a criticism of the profession than the model, although it is true that the presentation was obscure and gave the appearance of unreality.

Other academics have considered the problem since that time, and Neil Crosby has developed a model that has some of the elements of Wood's real value and which he describes as a real value/equated yield hybrid.

Crosby's starting point is Irving Fisher's classification of yield into three parts:

- time preference (i)
- expected inflation (d)
- risk (r).

We are all familiar with this, but it was Wood's contention that the return should be the interest rate for forgoing the capital, but excluding any additional return for the effects of future inflation, which he described as the real value or return, giving it the name 'interest risk-free yield' (IRFY), a combination of Fisher's 'i' and 'r'.

The approach takes the position that at each review a new rent can be negotiated that mirrors inflation over the antecedent period. That income would need to be serviced by a return for risk and the time element, but not for the element of future inflation. The exception is that as the initial rent is fixed and therefore inflation-prone, it is necessary to have an additional element of return to reinstate the purchasing power. The initial tranche of income would be valued at the equated yield as represented by present value of £1 in one year at $(1 + i)(1 + d)(1 + r) = 1/(1 + i)(1 + d)(1 + r)$, while subsequent tranches would be capitalized at the IRFY.

The IRFY (i) = $\dfrac{(1 + e)}{(1 + g)} - 1$

The purchasing power of the income is maintained by valuing at the equated yield, deferred by the appropriate period at a yield calculated from the above formula.

An example will show how the model operates and how it compares with the traditional term and reversion method.

Example 7.15

Value an office block recently let at £125 000 per annum on an FRI lease with upward only reviews at five-year intervals. The current market yield on an all-risk basis is 6.5 per cent, while the equated yield is 11 per cent.

Traditional method

Current rack-rental value	£ 125 000
YP perpetuity @ 6.5%	15.3846
Capital value	£1 923 077

By real value/equated yield hybrid approach

$$i = \frac{(1 + e)}{(1 + g)} - 1$$

$$\left[\frac{1.11}{1.0507}\right] - 1 \text{ equals} \qquad 0.05644$$

$$\text{which is} \qquad 5.644\%$$

Then valuation is obtained from:

Current rack-rental value   £125 000

$$\text{YP} = \text{YP 5 years @ } 11\% \times \frac{\text{YP perp @ 5.644\%}}{\text{YP 5 years @ 5.644\%}}$$

which is $3.6959 \times \dfrac{17.7179}{04.2554}$ $\dfrac{15.3884}{£1\,923\,550}$

Which gives the same result (subject to rounding approximations).

Short-form discounted cashflow

With continuing concerns expressed about the traditional term and reversion approach, the short-form discounted cashflow option has been proposed as an alternative.

The difference is that in the short form, the term income is discounted at the equated yield, as is the deferment of the reversion. In general terms and where rental growth is significant, the short form shows a higher capital value; in other words, it suggests that a term and reversion valuation based on the all-risks yield undervalues the investment. The longer the reversion, the more this approach seems to be at odds with market evidence; investors purchase property to share in income growth, and the longer the period for which the rent is fixed, the less like a property investment it appears to be. There may be an implication that investors purchase on the basis of rental growth but are more confident where the benefits are not too long delayed.

The implied rental growth is calculated from knowledge of the all-risks yield, the market or equated yield and the interval between rent reviews. It is then used to discount successive tranches of projected income at the equated yield rate. At the end of the period, the remainder of the capital value is calculated by treating it as a reversion and discounting using a growth-implicit yield.

Two aspects need further investigation. First, the holding period may be for any length of time but is usually set at between 15 and 30 years. A 15-year period would certainly seem the shortest period over which to reflect anticipated rental growth; a 30-year period, over which substantial growth is part of the attraction of the investment, is questionable. Changing needs may put the design and layout and possibly the location of the property at a disadvantage. Second, the all-risks yield used at the end of the holding period is unlikely to be the same as that derived from market sources at the time of the original valuation.

The following example takes a 20-year holding period as the longest over which a forecast of sustained rental growth is appropriate; it has been assumed that similar properties to the one being valued, but 20 years older, can be identified to give an appropriate all-risks exit yield.

Example 7.16

Value the freehold interest in modern office premises let on FRI terms with ten years unexpired, the current rent being £15 000 per annum. The current rental value is £21 000 per annum, assuming five-year reviews. Recent sales confirm an all-risks market yield of 6.5 per cent, while the market rate is 11 per cent. Compare a valuation by the traditional term and reversion approach with an assessment using the short-cut DCF model.

	Traditional valuation			
Term	Current rent per annum	£ 15 000		
	YP 10 years @ 6.5%	7.1888		£ 107 852
	Reversion to Rental value	£ 21 000		
	YP perpetuity @ 6.5% deferred 10 years	8.1958		£172 112
		total		£279 944

With an equated yield of 11%, the anticipated rental growth is 5.07%
Rent in 10 years' time will be £21 000 × 1.6400, which is £34 435

	Short form		
Term	Current rent per annum	£ 15 000	
	YP 10 years @ 11%	7.1888	£107 832
	Reversion to Rent in 10 years' time	£ 34 435	
	YP perpetuity @ 6.5%	15.3846	
		£529 769	
	PV £1 in 10 years @ 11%	0.3522	£186 585
			£294 417

Commentary

Both approaches value the current rent for the next 10 years (although at different discount rates). The current rental value is £21 000 per annum which is the figure assumed in the valuation of the reversionary part of the interest on the traditional basis. It should be assumed that, at that point, the rent negotiated would be on modern lease terms and in particular with a five-year review pattern. Note particularly that no attempt is made to project rental growth to find a possible rental value in 10 years' time.

The short form valuation assumes that the rental value will continue to grow and on this basis calculates what it is likely to be in 10 years' time.

The valuation then proceeds to value the first tranche at the equated yield rate of 11 per cent and the remainder in perpetuity without further explicit reflection of rental growth, at the all-risks yield rate. Note that the value of the reversion is deferred at the equated yield rate.

COMPETENCE IN VALUATION

The valuer aims for accuracy in valuation although it is well accepted that valuation is based on interpretation and opinion and that a valuation of an interest by one competent valuer is unlikely to be exactly the same as a valuation by another equally competent valuer, even though both valuations are likely to be within an acceptable range.

There is a good deal of case law regarding negligence in valuation and, while it is not intended to review it in a comprehensive way, it will be useful to extract the basic guidance available.

In an early case, the Court of Appeal held that, in the absence of an alternative explanation, gross overvaluation 'may be strong evidence either of negligence or of incompetence' (*Baxter v. F. W. Gapp & Co. Ltd (1938)*).

Many valuers believe that they can value to within 5 per cent; giving a range of 10 per cent. This belief is not borne out by the available information.

An experiment that drew much criticism was conducted by two actuaries, Hager and Lord, and reported in a joint paper delivered to a meeting of the Institute of Actuaries in 1985. The authors understood from informal discussions with valuers and others that the range of valuation for any particular property would be about 5 per cent either side of the 'correct' value.

They were therefore greatly surprised to find a much greater range when, seeking to test this hypothesis, they invited 10 surveyors 'who all have experience in asset valuation for pension funds, but do not necessarily have intimate knowledge of the locations chosen' each to value two properties. The first property, property A, with a control valuation of £725 000, was valued in the range £630 000–780 000, while property B, with a control valuation of £605 000, was valued at £450 000–655 000.

All but one of the valuations of property A were within the 10 per cent band. In property B, the two lowest valuations were outside the 10 per cent band. Whilst it would be wrong to draw any firm conclusions from the results, it is interesting to note that all three valuations outside the 10 per cent margin were below the relevant control valuations.

The experiment drew much criticism, not least from members of the surveying profession who complained that Hager and Lord were not valuers and did not understand the way in which valuations were carried out.

The first positive reaction to the Hager and Lord bombshell was for Drivers Jonas, a major national firm of surveyors, to sponsor research into valuation accuracy by Investment Property Databank (IPD). That role was taken over by the RICS in 2002, and in its first report in 2003 it concluded that the accuracy of valuation had improved but committed the institution to more research.

Included within its conclusions were the results of a set of descriptive tests in which it was found that:

- In 2002, 90 per cent of valuations were within 20 per cent either way of the target price, compared with only 59 per cent 20 years earlier.
- In the previous two years, 70 per cent of valuations were within 10 per cent either way of the target price, compared with only 39 per cent 20 years earlier.

The latest report (2006) discloses that some 86 per cent of sales were within 20 per cent of their valuation, although only 63 per cent were within 10 per cent of the valuation. A rising market and the time lag between valuation and sale are identified as responsible for the discrepancies. It must also be remembered that clients receive advice but may not follow that advice in its entirety when making the decision about the asking price. The comparison of valuation with price is not the same as the relationship of valuations of the same property.

There is now a great deal of emphasis on valuation standards and the RICS produces standards and guidance notes. The institution is a corporate member of the International Assets Valuations Standards Committee (IVSC) and the European Group of Valuers Associations (TEGOVA).

Another example where valuations differed notably centred on the holdings of the Oldham Estate, where one firm prepared a balance sheet valuation of £581 000 000, while another firm produced a valuation for the bidder of £436 000 000. The RICS acted quickly to investigate the valuations, but its claim that the valuations were undertaken within accepted professional standards was hardly reassuring given the institution's unwillingness to elaborate on its statement.

Where the question of accuracy was considered in a case before Megarry J, he concluded that he had to deal with the matter generally rather than with precise mathematics and to remember that valuation was an art rather than a science (*Violet Yorke Ltd v. Property Holding and Investment Trust Ltd (1968)*).

The same approach was confirmed some years later in a case before the Privy Council, which voiced its strong disapproval of any attempt to look at a valuation in too critical a fashion:

> In general their Lordships consider that it would be a disservice to the law and to litigants to encourage forensic attacks on valuations by experts where those attacks are based on textual criticisms more appropriate to the measured analysis of fiscal legislation.
> *A. Hudson Property Ltd v. Legal and General Life of Australia Ltd (1986)*

Further, the courts have recognized the difficulty in many cases of obtaining facts and confirming information provided by others, and have defended the need for a valuer to make assumptions:

> Often beyond certain well-founded facts so many imponderables confront the valuer that he is obliged to proceed on the basis of assumptions.
> *Singer & Friedlander Ltd v. John D. Wood and Co. (1977)*

In this case (which related to development land where the propensity for error may be greater), it was accepted by both sides that the permissible margin of error was generally 10 per cent either side of a figure that can be said to be the 'right' figure, while in exceptional circumstances that margin may extend to 15 per cent.

The same case underlined the importance attached to preparatory work and that the valuer should collect relevant information, analyse it and draw conclusions, but went on to acknowledge: 'The valuation of land by trained, competent and careful professional men is a task which rarely, if ever, admits of precise conclusion ... Valuation is an art, not a science. Pinpoint accuracy is not, therefore, expected.'

In the case of a development site, the parties agreed an acceptable margin of error would be within 17.5 per cent of the 'proper' figure (*Mount Banking Corporation Ltd v. Brian Cooper & Co. (1992)*).

Where surveyors were instructed to provide a 'franking' valuation of an extensive mixed industrial estate, the expert witnesses agreed that a margin of error of up to 20 per cent would be acceptable (*Arab Bank plc v. John D. Wood (Commercial) Ltd. (2000)*).

Unusual residential properties have been included in the 15 per cent margin: a residence converted from stabling (*Scotlife Homeloans (No. 2) Ltd v. Kenneth James & Co. (1995)*) and a house with a much larger plot than usual for the area (*Legal & General Mortgage Services Ltd v. HPC Professional Services Ltd (1997)*).

Other notable comments from court decisions include one to the effect that the valuer is concerned with market value (*Banque Bruxelles Lambert SA v. Eagle Star Insurance Co. Ltd (1994)*), although any adverse market trends should be drawn to the attention of the client (*Credit Agricole Personal Finance plc v. Murray (1995)*).

Three pieces of advice at a practical level would be:

- Analyse yields to within one-tenth or one-eighth per cent rather than one-quarter as happens frequently at the moment.
- Do not round off rents, yields, sub-totals and other information used in a valuation; rounding should be reserved to the end of the calculations.
- Report a rounded figure, so as not to imply a level of accuracy that does not and cannot exist.

SUMMARY

- The normal valuation process involves an interpretation of market activity.
- Each property is unique, so valuers will tend to reach different conclusions when applying market intelligence to specific valuation problems.
- The valuation is the value at a point in time.
- The valuer should be alert to what is proposed in the area and the extent to which it is likely to have an effect on the property being valued.
- Some valuers choose to value using direct market yields as evidence, carrying with them implications about future expected growth.
- Others prefer the DCF approach, where growth is projected and forms the basis of the valuation.
- Some values are constrained by specific statutory provisions.
- Any valuation must have regard to any contractual provisions, especially any that cannot be renegotiated.
- There is acceptance that valuation is not a precise process.

8 The residual method – the problem

The comparable and investment methods are the appropriate ones to use where transactional evidence of sales and lettings is readily available. However, it is often necessary to provide a valuation of undeveloped land or of land with obsolescent or otherwise unsuitable buildings incapable of producing an economic rent and where the site is ripe for development or redevelopment.

The passing rent, even where there is one, relates to the current use and probably reflects the deteriorating condition of the property. It will not assist in determining the value of the land for redevelopment (though it may be one factor in deciding whether development or redevelopment is viable at a particular time).

In essence the problem appears to be a simple one: the market is likely to relate the value of the land to the level of profitability of the proposed development. It may be that, in this quest for a value, recent sales of similar land for development will be of some assistance. For example, land for industrial development in any particular locality will tend to have a capital value within a known or identifiable range; current sale prices would reflect supply and demand related to the physical advantages and disabilities of the particular site. The best price in that market would be obtained for level sites that were well served by a local road network, close to a motorway junction and capable of easy connection to all main services.

Where there has been a recent transaction involving a site similar in all respects to the one under consideration, that is likely to prove the best evidence for the value of the site. But the chances of finding similar development site comparables are low. A difference in any material respect will tend to be reflected in a change in the unit value.

FACTORS THAT COULD AFFECT VALUE

- The geometric shape of the site, its dimensions and the road frontage for access and prominence.
- The net developable area in relation to the total site area, parts of which may be affected by planning conditions and/or restricted by service ducts, easements, overhead power cables and physical features such as steep slopes and flood plains.

- The geology and soil mechanics of the site, which may necessitate the use of a more expensive raft to cope with differential settlement or a foundation design that will accommodate heave in the case of heavy clay soils.
- The location of the site, which will be a major determinant; even adjoining sites can be subtly different.
- Planning permissions vary; they may contain conditions such as hours of business opening, onerous obligations with respect to on- and off-site works and services in the case of an outline planning permission, or may reserve a decision on certain matters pending submission of detailed proposals.
- The nature, extent and design of the proposed developments, even if both are industrial, may also be very different. One may be a high bay steel portal framed warehouse, the other a standard eaves height production unit of traditional brick and steel construction.

While not exhaustive, the list demonstrates the scope for differences of significance in the factors affecting value. As a result, any attempt to determine market value by comparison pricing on a unit basis is likely to prove both unreliable and inadequate.

Value is often expressed as a capital sum per unit of measurement, but the detail of the development, the size of the site and its access, the various expenditures involved and the potential for letting are likely to ensure that comparable information of the kind mentioned is not directly applicable and acts as no more than a rough check or possibly as a useful initial way of reporting the result.

THE BASIC APPROACH

The approach most often used in valuations of development land is the residual method. Based on as much information as is available, an indication of the land value is obtained by preparing a valuation of the proposed development on completion, deducting the estimated costs of carrying out the works, including fees and finance costs. As will be seen later, that surplus is not necessarily the value of the site. Table 8.1 sets out the process in simple terms.

Difficulties arise in the use of this method because the premises have not been built, plans and costings are tentative and rents are unlikely to have been negotiated.

A valuation is being placed on a completed development on the basis of information available on current rents and yields derived from an analysis of the current market, whereas the development may not be completed for two or three years, or even longer, by which time the market may have undergone considerable changes with a consequent effect on value.

Similarly, building costs or the cost of short-term finance, or both, may be affected by external events, while any delay in the development schedule (often incurred due to complications in obtaining planning permission, or interruptions due to events such as inclement weather or an unsuccessful marketing campaign)

Table 8.1 The residual method: the main components of the development process

Preliminaries	Construction
Appoint professional team	Contractor to build
Finalize plans	Marketing campaign
Obtain detailed planning permission	Agree rents and lease terms with tenants
Negotiate building contract	
Arrange short-term finance	
Appoint building contractor	
Arrange sale of completed development	

Post–construction
Finalize lettings
Fit out units (tenants' responsibility)
Complete sale of development when fully let
Repay short-term finance

is likely to have serious financial implications, especially when the cost of short term finance is high. So, while no value profile should give a false impression of its degree of accuracy or certainty, it behoves the valuer to take great care to ensure that all data used are based on the highest quality of information available.

The great advantage of the approach is that it mimics the way in which the market considers the problem: in other words, it is a market approach.

Where the land is already owned by the developer or the price has been agreed or is fixed, a variation of the method may be used. The known information could be used to determine the maximum amount available for building costs, the minimum net rent required from the completed works or the gross profit likely to result from the development. In each case, the objective is the same: to establish whether a development of the site is viable according to the criteria available (some, such as rents and yields, market-derived; some, such as the precise terms of any planning permission, externally proposed; some stipulated by the particular developer).

The criticisms

On many occasions, the Lands Tribunal has expressed serious reservations about the use of the method, which is often used by expert witnesses in cases of compulsory acquisition referred to the tribunal when the parties have been unable to agree the compensation to be paid.

In the following extracts from the decision of the tribunal in *Clinker & Ash Ltd v. Southern Gas Board (1965)* the remarks are directed primarily at cases where the market will not be tested because the land is needed for purposes for which compulsory powers are available. The normal operation of the market is first described and then contrasted with the situation before the tribunal, when only two parties are involved and the acquisition of the site is not exposed to the normal negotiating procedure.

From the viewpoint of a valuer who is retained by an intending vendor and who has therefore a responsibility to ensure that his client obtains not less than the full value of his land, there is a natural tendency to adopt somewhat full figures for the variables which together make up the completed value and/or to adopt somewhat conservative figures for the variables which together make up the development cost. Conversely, from the viewpoint of a valuer who is retained by an intending purchaser and who has therefore a responsibility to ensure that his client does not pay more than the full value of the land, there is a natural tendency to adopt somewhat conservative figures for the variables making up the completed value and/or somewhat full figures for the variables making up the development cost. At this point of divergence between the two valuers, however, the discipline of open market conditions intervenes, imposing external sanctions which are highly effective. The external sanction facing the valuer for the intending vendor is that, if his choice of figures for the variables should throw up too great a difference between completed valued and development cost, his client may well fail to find a purchaser at all because the calculated site value is above actual open market value. The external sanction facing the valuer for the intending purchaser is that, if his choice of figures for the variables should throw up too small a difference between completed value and development cost, his client may well fail to buy the land at all because the calculated site value is below actual open market value. It is a striking and unusual feature of a residual valuation that the validity of a site value arrived at by this method is dependent not so much on the accurate estimation of completed value and development cost, as on the achievement of a right balancing difference between these two. The achievement of this balance calls for delicate judgment but in open market conditions the fact that the residual method is (on the evidence) the one commonly or even usually used for the valuation of development sites, shows that it is potentially a precision valuation instrument. If there are two equally proficient valuers acting respectively for a willing vendor and a willing purchaser they would thus expect to agree on a price for the site in question, it being irrelevant for this purpose that one valuer may have arrived at the agreed site value by using figures for completed value and development cost differing substantially from those used by his colleague.

The tribunal then goes on to consider the use of the method in cases such as the one before it where there is to be no market transaction and which in their view presents dangers:

> When a residual valuation is prepared for arbitration purposes, however, the conditions are very different; the valuation is then a calculation made *in vacuo*; and although there may be a deemed open market there are no external sanctions acting as an incentive to the achievement of the delicate balance which I have described, because there is in effect a captive purchaser and a captive vendor. Thus there is no risk on the vendor's part of losing a sale by reason of the price advised by his expert being too high, nor is there any risk

on the purchaser's part of missing a buyer because the price advised by his expert is too low. Possibly as a side-effect of this absence of any external constraint, the natural tendency of the vendor's (or claimant's) valuer to adopt full figures when calculating developed value and conservative figures when calculating development cost almost invariably results (in the experience of the tribunal) in his putting forward an undependably high opinion of site value. Similarly the natural tendency of the purchaser's (or authority's) valuer to adopt conservative figures when calculating development cost almost always results in his putting forward an undependably low opinion of site value, and on occasion it may even throw up a minus site value. Having observed on so many occasions the working out of these tendencies in terms of widely conflicting 'valuations' the deep impression on the minds of the tribunal is that under arbitration conditions ... once valuers are let loose upon residual valuations, however honest the valuers and reasoned their arguments, they can prove almost anything. It is against this background and for this reason that the tribunal has reluctantly found itself compelled to reject the residual method when put forward as opinion evidence, unless there is no simpler method of valuation available.

The tribunal then considered the use of two substantial figures – the completed value of the development and the total cost of carrying out the work – to find the surplus available for site purchase and associated costs. The fact that the result is usually a smaller, albeit substantial, amount suggests that it is particularly sensitive to changes in the components used to find the figures first referred to. This observation is a further expression of the strong resistance to the use of the residual method where it is not exposed to the rigour and proof of the marketplace in a subsequent arm's length market transaction.

> In the form in which it is normally presented to the tribunal, the residual method for the valuation of a development site shows a site value which is thrown up as the difference figure between the estimated value of a completed development of the subject land, and the estimated cost of carrying out that development. The figures of completed value and development cost are usually both much greater than the difference between them, i.e. they are both much greater than the site value which is being sought. In the make-up of both the completed value and the development cost there are a number of variables; the appropriate figure to be adopted for each of these variables will depend on the viewpoint as well as on the knowledge and experience of the valuer; the choice from which each such figure may be made is a fairly wide one, varying from what may be termed 'conservative' to what may be termed 'full'; and within this range whatever figure may be adopted is 'correct' in the sense that it can be substantiated.

But if we consider the market composed of builders, developers, investors, speculators and others wishing to purchase land on which to erect a building for a particular purpose, they will quite naturally move towards the value of the site by a deductive process – that is, they will ask themselves the potential value of the completed development, and what the costs are of achieving that development including the return required for the effort, skill and risk involved in the transformation. The deduction of all costs from the value of the completed development will provide a guide to the maximum price they can afford to pay if their criteria are to be met.

It should be noted that the difference when expressed as the ultimate land value represents the maximum bid the developer can afford to make on the basis of the inputs used; other developers may arrive at higher or lower amounts reflecting levels of efficiency, costs of overheads, the individual firm's cost of borrowing or the strength of the developer's desire for that particular piece of land. Thus, strictly speaking, the residual method provides a 'calculation of worth' rather than a market value, as it is specific to the developer. The matter of worth versus market value is important. When developers are considering a bid, they need two points of reference: worth, being what they can afford to pay; and market value, being what they are likely to have to pay.

The unique calculations made by each interested party will result in a range of possible bids. The greater efficiency of a particular developer will ensure a higher residue, although not all of the savings will necessarily be made available for purchase of the site. A developer should be able to make the winning bid without sacrificing the whole of the increment attributable to the company's greater efficiency or other particular circumstances. The developer will wish to consider the likely bids of others and the benefits of purchasing the site. Except in auction situations, it is possible to make an initial offer to establish interest.

As an example, a developer may have calculated a residual of, say, £1 million for an office development site but would be ill-advised to make such an offer until the market price was known; the site may be placed on the market at £850 000 and it may be possible to obtain it for slightly less in the haggling of the market. Conversely, had the asking price been fixed at £1.1 million, a purchase at this price would have eaten into the developer's profits on the basis of the calculations made. Worth and value therefore set the parameters within which to bid, but where worth is less than market value the developer may decide to withdraw from the fray.

However, it would be wrong to conclude that time had been wasted: on the contrary, the exercise has served a purpose in indicating to the prospective developer that this particular site should be avoided unless there is a real prospect either of reducing costs or of intensifying development to increase the value of the completed project. This is one example of the ways in which development proposals are shaped and refined.

WIDESPREAD USE

The method is unique in that it is widely employed not only by valuers but also by developers and builders, i.e. 'the market'. At its simplest, the procedure is little more than a mental calculation, a process that may be acceptable where the site is a small uncomplicated one for a single unit of development and the calculation is made by someone having wide experience in the particular type of development envisaged and able to relate to a recent transaction involving a similar site.

It would not be prudent to try to assess larger or more complicated sites in the same way, particularly since the method has attracted much criticism due to its lack of sophistication. This need not be so. The robustness of the exercise can be tested statistically in various ways. There is room at least to look at a range of probabilities in rental values, building costs and costs of borrowing.

The next chapter will be devoted to ways in which the method is applied in a range of circumstances.

SUMMARY

- The residual method is used where there is potential for development, redevelopment or refurbishment.
- It enables a site value to be found without direct market evidence of value.
- The method simulates the approach taken by the market.
- Development is a complex activity, dependent upon a wide range of variables, thus making comparison a much less suitable framework for valuation.
- The rationale of the residual is that the investment value of the completed and fully let development, less the total construction and associated costs, leaves a surplus available with which the developer can finance the purchase of the land and the associated fees and costs.
- The input variables need to be well researched and carefully considered so that they do not have a knock-on effect throughout the calculation.
- The basic residual is highly flexible and can be used to determine the developer's maximum land bid, the profitability of a given scheme or indeed the amount of any input variable given a required land bid or profit figure.
- The basic residual can be criticized as being too simplistic in its approach to the timing of costs, including finance charges.
- The criticisms of the basic residual can be alleviated by the application of various cashflow models.

9 The residual method – the process

Having collected as much information as possible, the valuer will make use of it in determining the residual value of the site. (It is worth repeating that the available information may be used to produce other results; for example, the land may already be in the ownership of the developer or the price already fixed; the developer may wish to determine how much is available to spend on construction works or the minimum value of the completed development necessary to justify the payment of a certain land price.)

Figure 9.1 sets out the main variables to be included. Each item is considered in more detail below.

VALUE ON COMPLETION

The first step is to estimate the gross development value, being the capital value of the completed development. The lettable floor area is calculated and multiplied by the unit letting value to find the total rental value. A yield is then identified by the analysis of sale prices of other similar developments and experience of yields

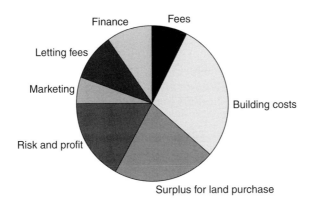

Figure 9.1 Typical constituents of gross value of the completed development

demanded by the market in general. This part of the assessment involves the investment method. The end-valuation is normally straightforward, envisaging a rack-rent capitalized in perpetuity for a freehold or, if leasehold, for the length of the ground lease available (a long period, probably 99 years or more probably 125 years unless the ground lease was granted some time ago, in which case the residue of the term will be available). Where the interest is leasehold, the frequency and nature of rent reviews, if any, of the head leasehold interest will be relevant to the yield, as will the effect of any unusual provisions in the lease that fetter the developer's ability to manage and operate the development in the best way. As it is assumed that the development will be sold on completion, the estimated costs of transfer are deducted from the final valuation.

DEVELOPMENT COSTS

In the initial stages of investigation, the developer is considering whether to make a bid for the site. It would not be cost-effective to prepare detailed plans at this stage, so it is likely that the information available to the valuer is provisional and sketchy. The outcome of the calculations will determine whether the site is likely to be of interest. If so, the developer will prepare more details of what is thought appropriate, which will enable the costings to be refined. The main elements of cost are now considered.

Building costs

The valuer should be concerned to earmark a sufficient sum for building costs and contingencies, especially since these have a 'knock-on' effect on other costs (for example, fees and finance). Too generous an allowance, on the other hand, may make the residual amount uncompetitive and result in failure to purchase the site. At this stage a good deal of information, expertise and experience is called for.

Details of building costs will eventually be determined when a contract is signed with the appointed contractor. Meanwhile, information will be gathered from a range of available sources. The developer might provide figures based on its own cost experience. Price books such as Spon's or Laxton's give a wide range of price information; the Building Cost Information Service (BCIS) provides a price subscription service under the auspices of the Royal Institution of Chartered Surveyors (RICS). An edited extract from *Spon's Architects' and Builders' Price Book* is given in Appendix A.

The proposed development will be measured, typically off plan, to reflect the total letting space. It is important for the measurements to conform to the standards laid down in the RICS *Code of Measuring Practice* (CMP).

The calculations should also include associated items such as demolition, site clearance and preparation where relevant. With an ever-growing emphasis on environment, costs may include the remediation of ground contamination as a prerequisite of planning permission. The valuer may be able to obtain the advice of a quantity surveyor; alternatively, the development team may have sufficient expertise to advise at this stage.

Professional fees and charges

Professional fees charged by architects, quantity surveyors and engineers might account for 12–15 per cent of the construction cost depending on the complexity of the building. Charges are levied by the local authority for planning applications and building regulation approval.

Site acquisition, the development itself and any sale of the completed scheme will involve fees, charges and expenses. The developer may agree to make a contribution to the professional fees of other parties (for example, as an incentive for prospective tenants) and will be responsible for any arrangement fee in respect of short-term finance. Stamp duty land tax will be payable. Where the developer is registered for VAT the amount paid can be ignored, as it will be recoverable. Charities and certain other groups are in a different position with regard to VAT, and it may be necessary to provide for all or some of the charges. An extended note on the impact on rents of VAT is included in Chapter 5.

The building contract

The building contract may be negotiated subject to a 'rise and fall' clause – the usual method whereby the client is exposed to any future changes in wage rates and material costs occurring after acceptance of the contractor's offer. Alternatively, the costs may be based on a 'fixed-price' contract, which is likely to be at a higher figure to compensate for the risks of future cost increases. In estimating the cost under this approach, the contractor will have regard to the length of the contract, the type of labour required, the ability to reserve materials for later delivery at a fixed price, the state of the company's order book, the general level of activity in the industry and a considered view of the wider economic indicators. Any time overrun is likely to provide grounds for the contractor to lodge an additional claim.

Contingencies

An allowance in the region of 5 per cent of the combined building costs and fees is made to cover the cost of unexpected items. The allowance may be higher where significant areas of cost have not been resolved in the early stages.

Short-term finance

Short-term finance is required to provide working capital to acquire the site, pay for professional services and meet interim and final certificates issued by the architect. The money is borrowed and interest paid for the period of the loan. Three distinct elements require finance: the *land acquisition*, the *construction*

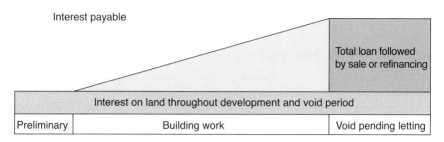

Interest payable

Total loan followed by sale or refinancing

Interest on land throughout development and void period

| Preliminary | Building work | Void pending letting |

Figure 9.2 Interest incurred during the stages of development (shaded)

(the building costs, professional fees and incidental expenses), and the *void period* (the period following completion of construction until the development is sold or refinanced).

Land acquisition

Borrowing for the land purchase will be a constant part of the total finance package to cover the site purchase, fees and associated preliminary costs. This is shown diagrammatically in Figure 9.2. Where there is a void period, the land element together with the construction costs will need to be financed for an uncertain period beyond physical completion of the works.

Construction

Borrowing for the construction phase will require progressive access to the total loan negotiated to enable payments to be made as costs are incurred.

Where it is possible to predict the rate of expenditure, a more sophisticated cashflow approach can be adopted. Otherwise, an accepted 'rule of thumb' is to calculate the total interest on the build costs and to assume that an average of half that amount will be borrowed over the period.

Void period

Once the development is completed the premises are available for occupation. But there is no certainty that a tenant or tenants will have been found at that stage and it is prudent to allow a void period by the end of which the building will be occupied. Many developments are speculative, in the belief that tenants can be found for the space created. Where possible, the developer will have already agreed sale terms with an institution or other investor to take effect once the development is complete and fully let.

Marketing the development will take place before construction is completed, but it is unlikely that all the space will be let by the time that it is ready for occupation.

Until this point is reached, short-term finance will continue to be required. At this stage, all the costs of land purchase, construction fees and other charges will need to be financed.

Where required, short-term finance is arranged through a bank or similar lender to meet the developer's need to borrow funds. The perception of property development is that of a higher risk activity. However the interest rate negotiated may also reflect the size of the loan, the proportion it bears to the total development cost, the existence of a pre-let or pre-sale and indeed the track record and financial muscle of the developer in question. In assessing a developer's best bid for the land, finance costs should be included in the residual even if a developer is acquiring the land and financing construction out of its own funds, so as to ensure that the opportunity cost of that money is reflected. In the event that there is strong competition for a site that the developer is particularly anxious to develop, the firm may choose to forgo some or all of the expected return on internal funding by adjusting the amount included for this item.

Where an investment fund is sufficiently interested in acquiring the completed development, it may well offer to provide the short-term finance as well.

Advertising and marketing

Advertising and marketing are crucial to the development. Publicity costs can be high and not entirely predictable; where lettings are not taking place as anticipated, it may be necessary to relaunch the marketing campaign and increase the budget by a significant amount. Expenditure may include newspaper, magazine, radio and television advertising, site boards, show building, negotiators on site, printing of brochures and other publicity material, press releases and general launch costs. The Internet provides a new advertising medium offering extensive cover. The allowance is best estimated by costing the proposed campaign and adding a margin; it would be inappropriate to calculate the cost as a percentage of the capital or rental value unless the developer had considerable experience with the particular type of development.

Agency fees

Agency fees will be incurred in the letting of units in the completed development and the investment sale of the tenanted entity. Where joint agents are appointed, the total costs are likely to be higher.

Fees will usually be expressed as a percentage of the estimated rental value used in the gross development value. Investment sale fees represent the purchaser's prospective costs of acquisition, to include an agent, solicitor and stamp duty, with investment sale fees amounting to some 6 per cent of the sale price. The purchaser deducts the sale fees from the gross development value to arrive at the net price to be paid.

Other costs

Other costs may be incurred in compensating outgoing tenants or in winning the consent of occupiers of adjoining property (for example, infringements of rights of light or allowing a crane to traverse the air space). All letting and sales transactions will be liable to stamp duty land tax.

Developer's profit

The developer's profit and risk will be shown as a cost of the development process. The amount charged may be a percentage of the capital value of the completed development or a percentage of the total costs involved, the allowance typically ranging from 10–25 per cent.

The level of profit and risk will be judged on the complexity of the proposal, the volatility of the outcome, the prestige of being associated with the particular development and the extent to which the profitability may be assured (possibly in part by a pre-arranged sale to a fund).

LAND VALUE

The gross amount available for land purchase is found by deducting the gross development costs from the net development value.

The remaining figure, the surplus, includes not only the cost of the site, but also fees, stamp duty land tax and finance charges in acquiring the site and holding it until the development is fully let and income-producing and thus capable of disposal in the investment market.

The net amount after deduction of costs is not a value in the strict sense, merely an indication of the maximum bid the particular developer can afford to offer for the site if the required returns are to be achieved on the basis of the information available.

The following example demonstrates the whole process in relation to a proposed office development. The accompanying commentary attempts a step-by-step explanation of the process.

Example 9.1 Office development: sum available for site purchase

A development company wishes to acquire a site in a provincial city where outline planning permission exists for the erection of an office block with a gross floor area of 1500 m². The accommodation will be on three floors served by two passenger lifts, and there will be surface car parking for 20 vehicles. There is a steady demand for office suites at rents in the region of £300/m². Short-term finance is available at 10 per cent, and the completed development could be sold to show a yield of 6 per cent. Construction costs are likely to be in the region of

£1000/m² over a building period of 12 months. It is anticipated that the whole of the building will be let within three months of completion of building works. Advise the company as to the maximum price it can afford to pay for the land on the basis that a profit of 15 per cent of the gross development value is required.

Gross development value on completion of scheme

Estimated rental value: 1200 square metres @ £300 psm	360 000	a
Years' purchase in perpetuity @ 6%	16.6667	b
Gross development value	6 000 000	c
Deduct costs of sale @ 4.5%	270 000	d
Net proceeds of sale	5 730 000	e

Gross development costs

Building costs: 1500 square metres @ £1000 psm	1 500 000		f
Contingencies @ 5% on building costs	75 000		g
Professional fees @ 15%	236 200		h
	1 811 250		i
Short-term finance @ 9%			
Construction costs (average 50%)	78 750		j
Fees (average of 75% for building period)	177 188		k
Total costs over void period (3 months)	40 753	296 691	l
		2 107 941	m
Lettings			
Agents' fees @ 10% of rents	36 000		n
Marketing campaign	15 000	51 000	o
Total cost of development		2 158 941	p
Developer's risk and profit			
15% of gross development value	900 000	3 058 941	q

Surplus available for site purchase (gross)	2 671 059	r
PV £1 1.25 years @ 9%	0.8979	s
	2 398 283	t
Deduct site acquisition costs @ 6%	135 752	u
Maximum available for site purchase (net)	£ 2 262 531	v

Commentary

The workings show a basic calculation that will indicate whether a scheme is likely to prove viable, based on the assumptions made. It should be borne in mind that the initial study is very approximate; for example, it will be noted that the planning permission exists only in outline and there is no firm building cost. The valuer will be making many assumptions about the design and quality of the building, which will affect rental values, yield and building costs. The outcome in terms of the value of the land will need to be treated with caution.

The following comments relate to the reference letter alongside each item in the worked example:

a   The rent is arrived at from a consideration of current market comparables but will not be determined until the scheme is complete, so is referred to as an estimated rental value. As offices are measured for letting purposes on a net internal area basis in accordance with the RICS *Code of Measuring Practice*, the given gross area of 1500 m² has been reduced, in this case by 20 per cent, to 1200 m². Detailed plans would allow the net figure to be calculated more accurately.

b   The yield is determined by reference to known comparables of similar new or modern office buildings in comparable locations. As the development site is of freehold tenure it has been valued in perpetuity.

c   The gross development value (GDV) is the best estimate of capital value of the completed development. There is an element of risk in the assessment since the building is not yet available and the market for office accommodation may change before leases are finalized. On the other hand, it makes no provision for any increase in rental value by the date of completion.

d   Sale costs are deducted on the assumption that the development will be sold on completion, once the units are fully let.

e   The net proceeds of sale take account of the purchaser's costs of acquisition.

f   Building costs are assessed as accurately as possible in the light of the information currently available. Useful sources of building cost information are *Spon's Architects' and Builders' Price Book* (see Appendix A) and the BCIS cost index published by the RICS. The cost of construction is based on the gross external area. Where more accurate costings are available from a quantity surveyor or other source, they should be used.

g   Contingencies are unforeseen costs incurred during the period of construction. The allowance depends on the uncertainties in any particular development but typically would be 3–5 per cent on the sum of construction costs and professional fees.

h   Professional fees include expenditure on the services of architect, town planner, quantity surveyor, mechanical and electrical engineers and others where required. A rule of thumb of 15 per cent can be used unless more accurate information is available.

i   The sub-total of the major cost items, being the figure upon which developers will assess their margin if they choose not to base it on the value of the completed development.

j   The developer will incur interest charges on building finance or, where internal funds are used, there will be an opportunity cost as the funds are not then available for use in other activities of the company. Experience suggests that the average amount borrowed over the building period will be one-half of the total estimated building cost, including contingencies. This example assumes that interest would be charged on an annual basis, although it is more likely that the interest would be calculated quarterly. In either case, the cost would be 'rolled up' and the interest paid in one sum from the proceeds of sale at the end of the project.

k   A substantial proportion of professional fees is payable during the early stages of construction, which is reflected in taking an average of 75 per cent of the total estimated costs.

l   In anticipation that the building will not be fully let on completion, finance will be required for a further period, here estimated at three months. At this point, the whole of the building cost will be financed or assumed to be financed. It should be noted that the void finance on the land element is dealt with separately under the heading 'surplus' in the residual. (See note 's' below).

m   A sub-total gives the sum of the construction and finance costs.

n   The building must be let to create a saleable investment. The services of a commercial property agent will incur a fee, usually based on a percentage of the first year's rent together with the payment of disbursements for advertising and the production of particulars or brochures. Commission at a rate of 10 per cent has been allowed.

o   An estimate of the costs of marketing is included.

p   The total cost of the development is now found.

q   The reward to the developer for profit and for assuming the risk of development may be assessed as a proportion of the development value or of the estimated total costs. In this case, 15 per cent of the GDV has been allowed. The provision will vary with the complexity of the development and the desire of the developer to be associated with it.

r   The gross development costs are deducted from the net proceeds of sale to show the surplus available to buy the site, fund the interest charges and pay the acquisition fees. At this stage, the amount includes interest and costs.

s   First, the interest incurred in acquiring the site and holding it until the investment is sold must be deducted, leaving the amount available for site purchase and costs.

t   This then provides the sum after interest has been paid for the whole of the holding period of 15 months.

u   Acquisition costs assessed at 4.5 per cent are calculated.

v   The remaining amount is what is referred to as the residual 'value'. In fact, it is not a value and it is certainly not the market value. Rather it is the maximum bid the developer can afford to make for the land given the assumptions made. These inputs will be unique to that developer and thus the residual can be considered a calculation of worth.

It is interesting to note that the site cost represents some two-fifths of the 'value' of the finished development. This is a useful comparable when looking at other office development opportunities.

The flexibility of the residual method can be demonstrated with reference to further examples that build on the method detailed above.

The first demonstrates the use of the residual to determine the viability of a development where the asking price of the site is available, for example where a developer is already in negotiation with a landowner and a purchase price has been provisionally agreed.

Example 9.2 Office development: profit available when site cost known

The facts are as given in Example 9.1 except that the land is available at the asking price of £2 250 000. The developer wishes to know whether the proposal would show a profit.

Gross development value on completion of scheme

Estimated rental value: 1200 square metres @ £300 psm		360 000	**a**
Years' purchase in perpetuity @ 6%		16.6667	**b**
Gross development value		6000 000	**c**
Deduct costs of sale @ 4.5%		270 000	**d**
Net proceeds of sale		5 730 000	**e**
Gross development costs			
Building costs: 1500 square metres			
@ £1000 psm		1500 000	**f**
Contingencies @ 5% on building costs		75 000	**g**
Professional fees @ 15%		236 250	**h**
		1 811 250	**i**
Short-term finance @ 9%			
Construction costs (average 50%)	78 750		**j**
Fees (average of 75% for building period)	177 188		**k**
Total costs over void period (3 months)	40 753	296 691	**l**
		2 107 941	**m**
Lettings			
Agents' fees @ 10% of rents	36 000		**n**
Marketing campaign	15 000	51 000	**o**
Total cost of development		2 158 941	**p**

Add price required for site	2 250 000			**i**
Acquisition charges @ 4.5%	101 250			**ii**
	2 351 250			
Add short-term interest				
charges @ 9%	271 266	2 622 516	4 781 457	**iii**
Profit available on basis of fixed price for site			948 543	

The previous calculations are adjusted to take account of the additional information.

Commentary

The calculations have been adjusted in particular to take account of the information given regarding the price of the land. The key difference in Example 9.2 is that the developer's profit figure has been omitted and in its place the land cost and associated expenses are included. By so doing the bottom line is no longer the residual land value but the amount available to reimburse the developer for risk and profit.

i   Here the known net land cost of £2 250 000 is inserted.
ii   The acquisition costs of 4.5 per cent are calculated.
iii   The interest charges on the loan to cover the cost of the land plus the acquisition fees are calculated over the whole of the 15-month development and letting period. Measured as a percentage of GDV it is 15.8%, rather more than specified by the developer in Example 9.1. On paper, the developer will make an additional £48 500 profit.

Both examples may be criticized for the assumptions necessarily made. The residuals both assume that most borrowings are evenly spread throughout the build period and that costs are incurred from day one, neither of which is true in practice.

A cashflow approach would enable the valuer to be more precise in the timing of cashflows that may have a significant effect on viability, especially where short-term interest rates are high or development periods long.

The next example will demonstrate the traditional residual approach to a warehouse development and then use the information in a more sensitive way in cashflow calculations.

Example 9.3 Warehouse development by traditional residual method

Value a cleared site for a warehouse development for which planning permission exists. You have collected the following information.

The gross floor area is 1000 m²; the building cost is £260/m² and the units are expected to let at £75/m².

It is anticipated that units with a total rental value of £68 250 per annum will be let and occupied immediately on completion of work in 12 months' time and that the remaining units will be let by the end of the next quarter, i.e. in 15 months.

The information is first used in a traditional residual calculation and then three cashflow approaches are used. It will be appreciated that the cashflow approaches give the opportunity to refine the inputs, particularly the timing of payments and receipts. Nevertheless, it should be borne in mind that very little of the information used is certain and that price increases, economic crises, building delays or increased materials and labour costs may occur and affect the results. However, the more careful use of time-sensitive information available removes some of the force of the criticism of this method.

Gross development value on completion of scheme

Estimated rental value: 910 square metres @ £75 psm		68 250
Years' purchase in perpetuity @ 8%		12.5000
Gross development value		853 125
Deduct costs of sale @ 4.5%		38 391
Net proceeds of sale		814 734
Gross development costs		
Building costs: 1000 square metres @ 260 psm	260 000	
Contingencies @ 5% on building costs	13 000	
Professional fees @ 15%	40 950	
	313 950	
Short-term finance @ 10%		
Construction costs (average 50%)	13 650	
Fees (average of 75% for building period)	3 071	
Total costs over void period (3 months)	6 825	
Letting		
Agent's fees and marketing, say 15%	10 238	
Total cost of development	347 734	
Developer's risk and profit		
15% of gross development value	127 969	475 703
Surplus available for site purchase (gross)		339 032
PV £1 1.25 years @ 10%		0.8877
		300 959
Deduct site acquisition costs @ 5.5%		15 690
Maximum available for site purchase (net)		285 269

The basic residual approach is insensitive to most timings. Other approaches offer a clearer account of when funds are committed and the charges incurred.

Three variations of the cashflow approach, using the same information as in Example 9.3, are now shown.

Example 9.4 Warehouse development: period-by-period cashflow approach

a	b	c	d	e		f	g	h
Period (3 months)	Item	Outgoing	Income	Outflow	Net flow	Capital from previous period	Interest @ 2.5%/qtr	Capital outstanding
		(£)	(£)	(£)	(£)	(£)	(£)	(£)
1	Building costs (15%)	40 950						
	Professional fees (75%)	30 712		71 662	71 662			71 662
2	Building costs (30%)	81 900						
	Fees/marketing (30%)	3 071		84 971	84 971	71 662	1 791.55	73 454
3	Building costs (25%)	68 250						
	Fees/marketing (70%)	7 167		75 417	75 417	84 971	2 124.28	87 095
4	Buildings costs (30%)	81 900		81 900				
	Professional fees (25%)	10 238		10 238	86 451	75 417	1 885.43	77 302
	Rent for one quarter		5 687					
5	Sale fees	38 391						
	Rent		11 375	27 016	27 016	86 451	2 161.28	88 612
								398 126

The period selected for demonstration purposes is quarterly over the 15 months or five quarters; a developer monitoring the progress of a scheme would expect a major financial report at least once a month and access to information on the flow of funds on a daily basis. A modern computer program would provide such information routinely; in doing so, some of the risk of development is removed, since there is much more opportunity to review costs and be alerted to problems at an early date.

Example 9.5 Net terminal value approach

Period	Net cashflow (£)	Interest @ 2.5% per quarter	Expenditure to completion (£)
1	71 662	1.1038	79 102
2	84 971	1.07689	91 504
3	75 417	1.05625	79 659
4	86 451	1.0250	88 612
5	27 016	1	27 016
		Total	365 893
	Developer's risk and profit		127 969
		Total	493 862

Leaving £320 872 for site purchase (gross)

Which enables the maximum site bid to be ascertained.

Expenditure in each charging period is debited with the total borrowing charges for the remainder of the development period and to that extent overstates the costs outstanding at any one time. As the example shows, Quarter 5 incurs no interest as it will be paid out of the proceeds of the investment sale. Quarter 4 incurs one quarter's interest at 2.25 per cent per quarter and so on, increasing with each period.

The example is a simple one but benefits from the cashflow technique. Where payments or receipts are irregular, particularly where the development consists of a number of units that may be let or sold at different times, the cashflow approach is the only way in which a reasonably accurate account of the proposal can be presented.

Example 9.6: Discounted cashflow approach

Period (quarters)	Net flow (£)	PV @ 2.5%	PV of cashflow (£)
1	71 662.00	0.9756	71 663
2	84 971.00	0.9518	84 972
3	75 417.00	0.9286	75 418
4	86 451.00	0.9060	86 452
5	27 016.00	0.8839	27 017
	Add risk and profit @ 15%		127 969
		Total	473 491
	Gross development value		853 125
	Current site value (less costs)		379 634

In this case all expenditure is discounted and deducted from a similarly discounted capital value, reflecting the elapsed time before the investment may be realized. The advantage of the period-by-period approach is that it shows the actual indebtedness at the end of each charging period, rather than the discounted cost and is therefore more helpful for budgeting purposes

The cashflow converts all costs into present day equivalents as they are incurred, allowing the site value to be deduced directly.

CONCLUSION

It should be noted that the flexibility of the residual approach allows it to be adapted to cope with much or little information. It can be used with small-scale development and expanded to cope with much larger phased development taking place over a number of years.

The availability of powerful computers or even hand-held personal calculators enables the valuer to carry out calculations as shown in the examples given and to track progress against forecasts. Commercial packages are available, but it is essential that they are used in the knowledge of what assumptions are built into the program.

For the student reader, the above alternative approaches may seem difficult to follow at first, but a very helpful way to achieve an understanding of the approaches is to devise a computer spreadsheet and compare the results with one calculated as previously described. Any program will need adaptation to reflect the peculiarities of the case in hand, and a thorough understanding of the workings of the program is required.

10 The profits principle

INTRODUCTION

This method, the fourth of the five described here, has long been known as the profits method or principle. There has been a tendency in recent years to adopt what is regarded as a more explicit title, namely the receipts and expenditure method, particularly where it is used to arrive at rating assessments.

In established trading positions in towns and cities where a variety of sales outlets trade side by side, the valuer will, in normal circumstances, judge rental levels by analysing the rents of premises that are let and applying the results to the unit to be valued. Allowances will be made for size, relative location within the trading area and lease variations, and for advantages and disadvantages over the referential material available.

The outcome should be a fairly close appreciation of the appropriate rental value, reflecting the opportunity for the tenant company to make a profit from its operations and for the landlord to receive a market rent. In other words, the market rental value is estimated by use of the comparative method described in Chapter 6. It is then only one step from formulating a capital value, should one be required. This route was fully explored in Chapter 7, where the investment method of valuation is described.

The valuer is primarily associated with the valuation of interests in land and buildings. However, certain types of businesses have a monopoly element that will have a strong influence on the trading results. It may be the building, the use to which it is put, the location or some combination of these features.

Businesses such as racecourses, theme parks, ski slopes (with hotels, restaurants and ski lifts) and casinos are sufficiently unusual to limit the scope for valuation by comparison. In the examples given, no product is sold as the main activity of the site; visitors pay to watch or take part in events and activities and the value is therefore likely to be directly related to takings. In some activities such as horse racing and skiing, the weather will also play a part in the level of activity.

Also, although not monopolies in the real sense of the word, public houses, filling stations, restaurants, leisure activities and similar types of use are notoriously difficult to value by comparison, and the profits principle is often applied by specialist valuers of these types of business.

DESCRIPTION

The profits principle is used primarily to establish a way in which businesses that exhibit some element of monopoly or uniqueness may be valued by reference to profits. There is an acceptance that they cannot be valued by the more usual methods because there are no similar premises or businesses with which they may be compared.

The Valuation Office Agency also tends to use the approach, in appropriate cases. It is regarded as more reliable than the contractor's test for the types of use described above in cases where the profit can be identified fairly readily.

THE VALUATION APPROACH

The basic approach consists of using the final accounts or other trading information to find the gross and net profits. Allowances are then made for interest on the tenant's capital employed in the business and remuneration for the management expertise. The remaining amount (referred to as 'the divisible balance') is then assumed to be available for an equitable division between the business trader and the owner. The balance is often apportioned equally but there is no reason for this to be so. The objective is to reflect the contribution of each to the success of the enterprise.

The reason for adjusting accounts in this way is to provide an adequate reward to the tenant for running the business, while properly reflecting the contribution of the landlord in providing the premises. Specialist valuers engaged in one or more of these areas will have a sound knowledge of the patterns of return, and this will enable them to make a fair allocation of the amount between the two parties.

Where possible, the valuer uses figures derived from recent accounts to measure the sums involved in achieving the annual turnover and the level of expenses incurred before gross and net profit figures are struck.

A well-established practice is to look at the accounts for the three most recent trading years. A series of accounts should give a more balanced indication of the general health of the business than could be obtained from a set of accounts for one financial year alone. The valuer is likely to be more influenced by accounts that provide results and trends over a period rather than by one year's accounts, which may not be representative.

The reasons given for any inability to produce accounts for the most recent year should be examined. Where the accounts appear to be subject to unreasonable delay, further enquiries are essential. It may be possible to gain some insight into the content of uncompleted accounts from the trader's accountant (but only of course with the trader's co-operation). There is no substitute for the audited and certified accounts if the valuer is to reflect the current activity in forming an overall judgement about the value of the business.

SITUATIONS WHERE THE PRINCIPLE MAY BE USED

The profits principle is used in a range of cases where any of the other approaches would be unlikely to be satisfactory. It is used by valuers specialising in, and with wide experience of, the particular type of business activity under consideration. Not only is there a need for an intimate knowledge of the type of business, but the valuer should be aware of the statutory provisions if the valuation is for rating or compulsory purchase purposes. Such specific applications of the profits principle are beyond the scope of this text and other sources should be consulted.

The valuer may be instructed to assess rental or capital values in four distinct situations:

- asset valuations (for inclusion in financial accounts showing property holdings as business assets);
- assessments for non-domestic rates;
- assessment of compensation on compulsory purchase;
- on sale as a going concern.

Asset valuations (for inclusion in financial accounts showing property holdings as business assets)

The valuation of property occupied and used in the course of business is strictly prescribed by financial reporting standards issued by the Assets Valuation Standards Committee (AVSC). The committee sets out the general principles to be observed in the preparation of asset valuations. It also makes recommendations and gives guidance, both of which are of crucial importance where the results will be publicly available.

Assessments for non-domestic rates

Where premises are to be assessed for rating (local taxation) purposes, the assessment will, in the majority of cases, be made on information drawn from comparable lettings of similar, nearby units. Where the property is of the type described in this chapter, the profits principle will be the most appropriate valuation method. In rating valuations, it is often referred to as the receipts and expenditure approach.

As previously stated, it may be used in valuations where the cashflow results from some form of monopoly attached to the property or location such that it is not possible to set the assessment by comparison. This situation may be brought about by legal requirements such as a licence to operate (e.g. casinos, betting office) or by the particular location of the business. Further, rating valuers may find the approach useful where there is no monopoly but a dearth of comparisons. Where the use of the premises provides a direct opportunity to make a profit, the approach is likely to provide a more realistic and reliable result than the contractor's test.

The complicated rules and regulations relate to the requirement that a hypothetical tenancy is to be assumed. The detailed application of the valuation process is beyond the scope of this text, but the following brief account contains some hint of the complexities of an assessment for rating.

The approach follows the profits basis as described above subject to any adjustment to reflect the extensive statute and case law related to non-domestic rating. So a hypothetical tenancy on a year-to-year basis is assumed, where the tenant is responsible for carrying out the repairs necessary to maintain the property in a state to command the rent. This provision may well differ from any actual liability imposed by a lease entered into by the tenant. It is also subject to the figure being related to the relevant valuation date (known as the antecedent valuation date). The latter is also referred to as the 'tone of the list', which is intended to ensure that all hereditaments have the same valuation date. An assessment taking account of these variations is the appropriate measure of the assessment. The valuation will include rateable items of plant and machinery provided by the tenant; whether plant and machinery are rateable is the subject of regulations made under statute law.

Examples include the assessment of hotels and filling stations. Hotel managements strive to develop a unique identity. The hotel may be known for its convenient setting close to the business area of a city combined with conference facilities; it may excel in the quality of its en suite facilities or it may occupy a site of some acres of lawns and gardens within a National Park. A valuer with a sound knowledge of the trade will have sufficient experience to rank the hotel and adjust for its special features. Similarly, each filling station exists to sell as much fuel as possible. But each location is unique. Sales depend primarily on location, but may be influenced by pricing, ease of access and facilities such as toilets. Recently, supermarket groups and others have shown an interest in developing express stores at the rear of filling station sites. It is claimed that each activity helps the other.

Assessment of compensation on compulsory purchase

This category is for the assessment of compensation for compulsory purchase for the acquisition of the whole or any part of any premises to include loss of business where appropriate.

The types of business previously described will be valued on the profits principle unless another method is considered more appropriate in any particular circumstances. It is the tenant's responsibility to minimise any loss and to relocate if feasible. The necessary expenses of relocation and temporary loss of business are proper heads of compensation. The valuer will assess the value of the business premises and, where trading prospects are affected or extinguished, any loss of profit and forced sale of business assets.

On sale as a going concern

When a business is for sale in the open market as a going concern there may be a goodwill element. The price will then reflect not only the intrinsic value of the physical premises but the benefit of acquiring an established business with its clients or customers who may be expected to continue to patronise the business after sale to the new owner.

It may be appropriate to apply a variation of the profits principle to such premises, especially where, although not a monopoly, the business may be of a form unique to the immediate area. An example would be a village general stores and sub-post office, typically the only business of its type in the locality and incorporating both a retail area with storage and living accommodation. One approach would be to value the business in two parts. First, the rental value is capitalised or experience used to go direct to capital value based on the residential market in the vicinity, adjusting for the retail element. Next, the goodwill is assessed by taking the net profit (excluding any rent or rental value of the retail part) and applying a modest years' purchase, typically between one and five.

When the business is subject to a lease, the profit rent is assessed and capitalised for the remainder of the unexpired term (or period to the next review if shorter). The premium so calculated reflects the benefit of a rent below current rental value and the security provided to business tenants by statute (unless excluded by agreement). Should greater security be sought, one or both parties may apply to the landlord to grant a new lease to the purchaser of the business. Such a request is likely to be met with proposals to update the lease terms and probably negotiate a new rent level. Any changes introduced by a new lease are likely to affect the level of any premium.

At one time, this type of business and living accommodation in a popular village would attract considerable interest. The opportunity to work from home with a post office salary as a base income together with the profit made from the retail activities – such as newspapers and magazines, groceries, greengroceries and household goods – would make for a convenient and profitable enterprise.

However, the demise of many sub-post offices in recent times and an uncertain future for those remaining has made the sale of such businesses problematic. Today, the main influences are likely to be the quality of the living accommodation, its location in a favoured village and the potential if the shop portion becomes uneconomic and is closed at some time in the future.

LIMITATIONS

Valuers dealing with business transfers (whether the business is offered with or without the freehold interest of the premises), rating assessments or claims for compensation for compulsory purchase will need to have access to financial accounts. But they should always keep in mind that they are not accountants and should not trespass beyond their understanding and ability. It is always possible and sometimes prudent to seek information or advice from the client's accountant or another conversant with the type of business carried on. This would particularly be the case where the business is complex or has a large turnover. In compulsory purchase cases, the cost of specialist advice would form part of the claim for compensation.

Potential

A prospective purchaser may hold the view that current profits could be improved substantially by better management, improved financial controls, or the incorporation of other sales lines or improvements to the business. To the extent that any potential would also be recognized by other prospective purchasers, the ability to create additional profits will be taken into account in any offers made by interested parties. Such aspects should be reflected as an intrinsic part of the value of the business, although the eventual purchaser would expect some benefit for the effort and cost of harnessing the potential. In fact, there may be a perfectly good reason why the vendor has not implemented the change (or perhaps did so in the past and found it unsuccessful). The purchaser will be taking the risk of failure associated with any change or new idea.

Special features

An important part of the value of some types of business to which the profits principle may be applied is the personality of the owner or operator. It is well known that a business fronted by a celebrity or character is likely to experience an increase in custom. For example, a famous chef will be of immense value to the reputation of a restaurant. Ex-boxers are often very successful public house landlords. The same may apply to experts; a car tuning station run by a former racing driver or a sports centre supervised by an Olympic athlete are likely to attract additional business by association with the well-known name. That element of value is unique and cannot be transferred. Another well-known person could attract his or her own following, but it would be wrong to pay a price based either on the reputation of the vendor or on the expectations of the drawing power of a particular new owner.

Conversely, it is sometimes found that a business has been neglected by an owner who is financially restricted, ill, lazy or incompetent. Such a business may be judged capable of substantial improvement given competent and unfettered management. To the extent that it can be judged, some allowance may be made for such a prospect, although again the prospective purchaser would expect to negotiate a price that acknowledged the effort, time and risk associated with any upgrading.

BUSINESS STRUCTURES

There are two main types of business structure: personally owned and operated businesses; and limited liability companies.

Small businesses may be operated by the sole owner or by two or more (normally not exceeding 20) owners trading as a partnership. Each and every partner is responsible for the debts of the firm. Finance is raised from the owners of the firm and from banks and private loans, often secured on some or all of the business assets and sometimes personally guaranteed by one or more of the owners.

Larger organisations tend to trade as limited companies. The extent of the liability of shareholders in a company is limited to the amount invested or promised to be invested; they have no further liability or call on their other assets.

Public and private limited companies are regulated by the Companies Acts. There is no bar to the small individual trader forming a company, thereby limiting its liability as described.

Another advantage of trading as a limited company is that initial capital and any further funds necessary may be raised by the issue of shares or debentures. The shares of public companies are traded on the Stock Exchange or, in the case of smaller limited companies, on the Alternative Investment Market.

There are different classes of share: preference shares carrying a fixed rate of interest and, the main class, ordinary shares. Ordinary shareholders will receive a variable dividend based on the profitability of the company. Preference shares take precedence in the event of liquidation. Both classes normally enjoy voting rights.

In the case of larger businesses, particularly those incorporated as companies, it is likely that any instructions to the valuer will be received from the accountant or financial team advising the company. In that event, the valuer will provide a report dealing with the limited brief. The report will include caveats to ensure that it is not used for any other than the stated purpose and that it should not be quoted from or shown to third parties without the prior approval of the valuer.

THE USE OF ACCOUNTS

Before preparing the final accounts a trial balance is extracted from the financial records. The purpose is to confirm the arithmetical accuracy of the double-entry system of bookkeeping.

The final accounts

At the end of each accounting period, normally of one year's duration, all financial transactions are brought together in a set of accounts that show the gross and net profits for the trading year and the capital assets of the business at the year end.

The final accounts comprise the trading account and the profit and loss account, which record the result of the year's trading. The trading account identifies the primary costs and receipts of the business to show a balance; where the receipts exceed the cost the amount shown is the gross profit. The profit and loss account details the remaining costs of running the business to show the net profit. In businesses engaged in manufacturing, the two accounts are usually prepared separately, otherwise they may be combined. The balance sheet shows the position of the business at the year-end in terms of its capital structure.

The set of accounts for the last complete financial year will give some indication of the health of the business. Individual components can be used to measure various aspects of the profitability of the business and various recognised ratio or performance indicators assist in this process. A more thorough picture of the business will be gained by analysing the accounts for the previous three financial years and observing trends.

The trading account

The trading account shows the gross profit, being the difference between the cost and the selling price of the stock including any carriage costs. To ensure a realistic figure, provision must be made to include the book value of the stock at the beginning and the end of the trading period.

Stock acquired for resale but not yet sold is normally recorded at cost price (but stock that has no market because it has been superseded or outlived its shelf-life would not be included except at net realizable value or scrap value, if any). Practice will vary according to the type of business and the accounting conventions employed.

Where a substantial amount of work is undertaken to convert raw materials into a component or finished goods, it is usual to include the cost of the work in the stock value. But it would be misleading to include the goods at their proposed sale price, thereby anticipating the expected profit on sale.

Debtors are recorded as assets unless bad debts are anticipated. Payment of an outstanding account will reduce the total figure attributed to debtors and increase the bank balance by a similar amount.

The bank balance recorded shows the amount held in the bank for normal trading purposes; the firm would not expect to keep more than is necessary for day-to-day operation because it should be better employed in the business.

Businesses supplying services rather than goods will prepare only profit and loss accounts, deducting all expenses from the income. Examples of such businesses are a solicitor's practice or a travel agency. Other businesses may choose to combine the trading and profit and loss accounts.

The profit and loss account

The profit and loss account shows the running costs of the business, which are deducted from the gross profit found in the trading account to show the true profitability of the business (the net profit).

The balance sheet

The balance sheet displays all the assets of the business including any profit or loss not distributed from previous accounting periods. The total amount shown is the theoretical value of the business – the amount available should the assets be sold. A business in sound heart would be expected to realize more than is shown for several reasons.

First, there is a tendency to write down assets for depreciation (and tax) purposes, regardless of their true value in the business or the market. After some time, many of the more durable assets of a business will have a nil book value although remaining in operational use. Second, purchasers would consider making a payment for goodwill, which may be described as the opportunity to continue trading with the company's current customers and acquiring new customers attracted by the firm's name, trademarks and reputation.

Numerical examples of the accounts described above are contained in the following chapter.

THE DIVISIBLE BALANCE

The divisible balance is an approach for apportioning the earning capacity of a specialized building that cannot be valued by one of the more conventional methods.

Information is gathered to enable the net profit to be identified. Any borrowings are added back and the tenant's capital employed in the business is reflected at an appropriate rate of interest. This adjusted net profit is the amount available to reward the tenant company for its entrepreneurial activities and to compensate the landlord for making the premises available. The balance is often divided equally, but each case must be considered on its merits. For example, because of its location or other peculiarities, the building may be difficult to let but is now occupied by a tenant who has exploited its potential for the business carried on. This situation would suggest that the tenant should keep a larger proportion of the available balance, say 60 per cent. A numerical example is provided in the following chapter.

ACCOUNTING STANDARDS

The context in which valuations are provided and used is of prime importance in the case of asset valuation. The valuation of fixed assets for financial statements is subject to the rules, practice statements and guidance contained in the *RICS Appraisal and Valuation Standards (the Red Book)*.

The main purpose of the rules is to ensure that any valuation is clear and unambiguous. They are designed to prevent the inclusion of values that are inappropriate and may therefore be misleading, intentionally or otherwise. Historically, at least in part, they were intended to deter the practice of asset stripping where a business was acquired on the basis of its going concern value but then closed down and the land ownership broken up and developed or sold for its best use. Effectively, this practice deprived owners or shareholders of part of the intrinsic value of the business.

The rules and requirements are extensive and detailed and the topic is beyond the scope of this book. However, the following summary is provided to alert the reader to the significance of providing property values for inclusion in accounts.

In the majority of cases, the capital value of a property asset for accounting purposes is, in broad terms, the open market value with vacant possession disregarding potential alternative uses and any other factors likely to affect the market value. This is termed the existing use value. It may differ from the market value where there are statutory or contractual limitations, where the site is overdeveloped or where it has old buildings.

The valuer is required to draw attention to the valuation being subject to the adequate profitability of the business, a matter for determination by the directors.

Specialised property in use in the business will be valued on the depreciated replacement cost basis (described in Chapter 12). Again, it is the responsibility of the directors to determine whether the business is sufficiently profitable to support the value at the level reported, or whether it should be included at some reduced figure.

The open market value basis applies only to land and buildings held for investment or where they are surplus to requirements.

The directors of a company are by law required to disclose the market value if it differs materially from the existing use value.

TURNOVER RENTS

Another approach to providing a fair rent for an income-producing property where the rental value is difficult to determine for one reason or another is to base the rent on the turnover of the business. In theory, the rent for the types of unusual properties discussed earlier in this chapter where there is little or no comparable evidence could be set in this way. There is little evidence of such arrangements, although undoubtedly they exist. Where it has become established is in the fixing of rents in large modern shopping centres where there is little or no direct comparable evidence, particularly where the developer identifies a substantial budget for promotions and other initiatives to launch the new centre and to enable it to become established quickly.

The approach is a logical one because rent is one of the expenses to be paid out of total takings: the landlord cannot expect too great a share of the profit if the tenant is to continue in business, while the tenant will have the comfort of knowing that, in the early stages at least, there is some understanding of the time taken to develop the potential of the business.

SUMMARY

- Some commercial properties are unusual in style or location.
- Comparison is not available, making another valuation approach necessary.
- The essential element in many commercial properties is the opportunity to make a profit.
- The profits principle looks at actual results and or potential trading prospects.
- From the trading prospects, a notional surplus is found that is then divided between the owner of the property and the trader in what appears to be fair proportions.

11 The profits principle – final accounts

This chapter sets out a simplified set of final accounts for an imaginary business, which is then used to demonstrate the way in which the information may be used in the valuation process.

The business deals in the sale of computers and consumables and includes a service facility.

The trading account is set out in Table 11.1. The profit and loss account is set out in Table 11.2. The balance sheet is set out in Table 11.3. The divisible balance is set out in Table 11.4.

THE TREATMENT OF ASSETS IN THE FINAL ACCOUNTS

Fixed assets include both tangible and intangible assets and investments: tangible assets include land and buildings, plant and machinery, fixtures, fittings, tools and equipment; intangible assets represent the company's view of the value of its goodwill, patents, licences, trade marks and similar rights and assets. There is no current uniform practice in relation to goodwill, but there should be no objection to its inclusion where there is some evidence that a value exists or where it has been purchased from a previous owner.

The net current assets are represented by stock, debtors and cash less the total amount owed by the firm to short-term creditors, such as trade creditors, any overdrafts from the bank and outstanding tax owed to the Inland Revenue.

PERFORMANCE INDICATORS

The final accounts for the year show the gross and net profit together with the assets, and give a general impression of the health of the business.

Further information can be gleaned by looking at the strengths and weaknesses of the business. The final accounts can be analysed in a number of ways, and where more than one year's accounts are available, trends can be deduced. There are accepted ratios and measures, some of which may be appropriate in assessing the underlying quality of the business under consideration. Some of the more useful are given over the page.

Table 11.1 Final accounts: the trading account

Final accounts for the HILO Partnership for the year ended 31 December 2007
Trading account

		£	£
Sales			343 045
Less	Returns	10 500	
	Discounts given	15 454	
			25 954
			317 091
Cost of sales			
	Opening stock at 1 January 2007	35 670	
	Purchases	123 052	
	Discounts received	– 1 485	
	Closing stock at 31 December 2007	– 44 500	
			112 737
Gross profit			204 354

Table 11.2 Final accounts: the profit and loss account

Profit and loss account

	£	£
Gross profit		204 354
Interest received		4 750
		209 104
Less expenses		
Distribution costs	20 100	
Administration expenses	72 526	
Bank charges and interest	3 124	
		95 750
Net profit		113 354

Table 11.3 Final accounts: the balance sheet

Balance sheet

Fixed assets	£ Cost	£ Depreciation	£ Net book value
Diagnostic equipment	16 500	1 500	15 000
Fixtures and fittings	35 000	4 250	30 750
Fleet vehicles	62 500	12 500	50 000
	114 000	18 250	95 750
Current assets			
Stock		44 500	
Debtors		15 250	
Cash at bank and in hand		17 255	
		77 005	
Current liabilities			
Trade creditors and accruals		43 725	
Other liabilities		13 252	
		56 977	
Net current assets			20 028
Net assets			115 778
Capital accounts			
Hi		68 970	
Lo		33 393	
			102 363
Current accounts			
Hi		7 700	
Lo		5 715	
			13 415
			115 778

Table 11.4 The divisible balance

To find divisible balance:		£
	Net profit	113 354
	Return on capital employed at 8%	9 262
	Divisible balance	104 092
Shared	60% to trader	62 455
	40% to freeholder	41 637
		104 092

Profitability

$$\text{Return on capital employed} = \frac{\text{profit before interest and tax}}{\text{total net assets}}$$

The profits of a business are related to the capital employed by the owner. The level of return would be expected to exceed that obtainable by depositing the capital in a relatively safe investment such as a building society account. The owner would expect a significantly higher return for the uncertainty and additional risk assumed in employing the capital in a business.

$$\text{Return on net assets employed} = \frac{\text{net profit}}{\text{net assets}} \times 100$$

This ratio measures profitability to the value of the net assets employed. The net assets are made up of the total of fixed and current assets less any current liabilities. The result may well understate the return as fixed assets are likely to be shown in the accounts at purchase cost less depreciation, which may well not be an accurate reflection of their current value.

$$\text{Gross profit margin} = \frac{\text{gross profit}}{\text{sales}} \times 100$$

There is likely to be an industry norm, but overstocking and subsequent sales may erode the margin. The gross profit needs to be sufficient to cover all the expenses of running the business. It follows that a high turnover may offer the opportunity to lower unit prices or to offer discounts to some customers.

$$\text{Net profit margin} = \frac{\text{net profit}}{\text{sales}} \times 100$$

Unlike the gross profit margin, the net profit margin will be important in throwing light on the efficiency of the business, and is a measure of particular importance where information on the performance of other similar businesses is known.

$$\text{Expenses related to sales} = \frac{\text{selling expenses}}{\text{sales}} \times 100$$

This figure will disclose the expenditure necessary to achieve the level of sales recorded. The owner may then be able to consider whether further sales could be achieved without increasing the resources used or whether there is a more efficient way of achieving the sales and at a lower cost.

Liquidity

$$\text{Working capital ratio} = \frac{\text{current assets}}{\text{current liabilities}}$$

This is a measure of the relationship between current assets and current liabilities. Bear in mind that a relationship of less than 2:1 may suggest instability. Too high a ratio should be investigated as it may suggest inefficiency.

$$\text{Liquid ratio} = \frac{\text{current assets (excluding stock)}}{\text{current liabilities}}$$

This is also referred to as the quick ratio or acid test. Ideally there should be an approximate equivalence between liquid assets and current liabilities. Stock is excluded as an asset as, although a current asset, it is the most illiquid. Should the current liabilities exceed the current assets by any significant amount, the firm could be in difficulty and under pressure to liquidate a large debt.

$$\text{Interest cover} = \frac{\text{net profit before interest and tax}}{\text{interest payable}}$$

Too low a ratio would indicate that the company is being stretched financially. Most firms carry some debt, but it is an expense that needs to be monitored closely, bearing in mind that most interest rates are variable and any increase is likely to occur when trading conditions have, or are about to, slow down.

Asset utilisation

$$\text{Stock turnover ratio} = \frac{\text{cost of goods sold}}{\text{average stock}}$$

This gives an indication of the average length of time that stock is held. The result will depend on the type of stock held; it will be shortest where stock has a short shelf-life, when it may be only a matter of days.

$$\text{Debtors' collection periods} = \frac{\text{closing trade debtors} \times 365}{\text{turnover}}$$

This gives the rate at which accounts are paid. A retail shop will have few, if any, credit customers, whereas a builders' merchant catering for the trade will transact most of its business on credit.

$$\text{Creditors' collection periods} = \frac{\text{closing trade creditors} \times 365}{\text{purchasers}}$$

A measure of how quickly accounts for goods and services are paid. A delay in making payments can be a cheap source of temporary finance but constant abuse of the credit facility will be noticed by suppliers and may result in a different level of treatment; special lines at advantageous prices may be reserved for the better customers.

$$\text{Asset turnover ratio} = \frac{\text{sales}}{\text{net assets}}$$

This is a measurement of the efficiency in the use of assets to achieve sales. This ratio can be used in showing the trend from year to year. The ratio will of course vary greatly from one business to another. Food stores catering for weekly needs and requiring relatively few fixed assets will have a high figure, whereas any manufacturing company, with a slower turnover and a greater investment in assets, cannot be expected to be as high. Nevertheless, it is a useful comparison between similar firms.

Companies employ measures such as dividend, earnings and price/earnings ratios, dividend cover and capital gearing, which serve a similar purpose to the indicators set out above.

Specialist valuers active in the use of the profits principle for the valuation of this class of asset will be familiar with the industry norm for the main ratios and would be put on notice if the results departed far from that accepted level.

Readers may wish to look at the sample accounts provided in terms of some of the ratios discussed above.

NUMERICAL EXAMPLES

Because of the statutory provisions overlying certain applications, some of the following examples are in outline only and do not reflect the considerations necessary in adapting the principle to that particular purpose.

Two of the following examples include summaries of actual cases. They may show Imperial measures and out-of-date unit values but their interest lies in the way in which the valuation problem was resolved. They remain relevant to the way in which the approach is applied in practice.

Example 11.1

Petrol filling station as reported in *Payne v. Kent County Council (1968).*

The Lands Tribunal member considered a compensation claim following compulsory acquisition of the petrol filling station, workshops and café adjacent to a bypass within a quarter of a mile of junction 3 of the M2.

The tribunal found itself adjudicating on petrol sales estimates of between 1 100 000 and 1 200 000 gallons and capital values of £500 000 to £1 135 000 claimed by the respective parties. The tribunal arrived at a figure of £750 000, made up as follows.

Estimated petrol sales in gallons per annum	1 500 000
	£
Rental value at 3.5p per gallon	52 500
Years' purchase perpetuity @ 6%	16.67
	875 000
Less cost of development	175 000
	700 000
Add to reflect opportunity to provide small catering unit	50 000
Compensation awarded	**750 000**

Commentary

As would be expected, the major part of the value is attributed to the favourable location of the filling station.

The petrol throughput and an estimated rental value per gallon are used to determine an annual rent, which is then capitalized. A lump sum is added to reflect the possibility that a small catering operation would be possible.

The valuation is of the simplest form but reflects the way in which the market would be likely to approach its assessment of capital value.

Example 11.2

A Lands Tribunal decision on the value of a petrol filling station acquired for a local road scheme (*Mayplace Garage (Bexleyheath) Ltd v. Bexley London Borough Council (1988)*).

The matter was complicated because two sites were involved, the filling station site in the ownership of the company and an adjoining site personally owned by the managing director of that company.

Evidence was given that the station was outmoded in layout and appearance, and in disrepair. The tribunal accepted that the sites could be merged for the purposes of assessing compensation.

The tribunal assessed the capital value in the following way.

		£
Estimated annual sales volume in gallons	750 000	
Capital value at 0.75p per gallon		562 500
Less estimated cost of redevelopment		200 000
Capital value (free of oil company supply agreement)		**362 500**

Commentary

Another simple valuation but it achieves the objective of relating the estimated throughput to the unit capital value available in the market.

The anticipated volume of petrol sales was used, but in this case the unit figure was related directly to capital value. The sales were those anticipated for a newly developed filling station and the estimated cost of new buildings.

The value of the site was assessed assuming it to be free of tie to any oil company.

Example 11.3

The business in this example is a sub-post office and general store, consisting of modern shop premises at the end of a terrace in a busy local shopping area some three miles from the city centre. The first floor accommodation is used for storage, apart from one room used as an office. The Post Office salary is £12 500 p.a., while the store is said to have a turnover of £1500 per week.

The business is for sale as a going concern, together with the freehold; the head postmaster is prepared to transfer the appointment to a suitable applicant.

The accounts have not been prepared by an accountant and are not sufficiently detailed to convey a full account of the operation. The accommodation on the first floor is not adequate for conversion to a flat.

There is no numerical example as the outcome is discussed in the text.

Commentary

It is unlikely that the profits principle would be appropriate in this case. The valuer would look at information about the open market value of similar shops in the same terrace or similar trading localities in the area. The general store and the sub-post office are complementary: they support each other. Loss of the appointment as sub-postmaster must be considered. It is unlikely that the head postmaster will be in a position to predict the future of the appointment as decisions on closure are made centrally and there have been many closures in the recent past. The future of general stores is also uncertain as they are increasingly under pressure. Supermarkets are now expanding into neighbourhoods and opening smaller convenience stores, often associated with filling stations. The ability to park for a limited time is an important feature of such developments.

The valuer should be able to value the premises and the business; a fairly modest sum could then be added to reflect the benefit of the Post Office salary, but bearing in mind its vulnerability. At one time, such an opportunity would be valued at up to five times the yearly salary, but any addition in this case would be much lower reflecting the realities of the situation.

No doubt the value of the vacant shop can be related to other transactions in the area relating to similar shop premises. The post office appointment represents a quasi-monopoly but the distribution of such facilities is under active consideration and there is no guarantee that the sub-post office will continue at this address. Closure of the post office side of the business would be likely to result in a reduced turnover from the general store.

Example 11.4

A 'free' public house occupies a prominent corner site in the retail centre of a prosperous market town. In addition to the sales of alcoholic drinks, there are two gaming machines, and bar snacks are served at lunchtime. The latest accounts show receipts of £475 000, showing a real increase on previous years.

The profits valuation is made on the basis of the ability of the building in its present use as a public house to produce a profit and may therefore give some indication of value.

A. Capital value based on financial information

		£
Net profit (including interest on short-term loans)		125 000
Add back interest paid		3 760
Adjusted net profit		128 760
Deduct:		
Interest on tenant's capital, comprising:		
a. furniture, fixtures and fittings	14 500	
b. stock value	8 000	
c. working capital	25 000	
	47 500	
at 8%		3 800
Adjusted net profit		124 960
Years' purchase based on accounts information		5
Capital value		624 800

(Equivalent to a years' purchase of approximately 1.3 of gross receipts.)

B. Finding the divisible balance from net profit

Check figures by using divisible balance to find rental value, then capitalize

Adjusted net profit (as above)	124 960
Tenant's share, say 60%	74 976
Balance available for rent	49 984
Years' purchase at 8%	12.5
	624 800

C. From recent enquiries you have received, you are aware that an international coffee house chain is anxious to find premises in the town. There is a good prospect that these premises would be suitable given their situation. The company is willing to consider a rent of up to £100 000 per annum. The rental value, in their present condition of the premises with the tenant responsible for shop fitting works, is in the region of £75 000 per annum. The investment, when let, would sell readily to one of the funds. No doubt your client will wish you to explore this opportunity.

Commentary

Sales of beer and other liquids are reduced to equivalent barrels, which simplifies assessment of the contribution to profit of each type of drink. Additions are then made for other trade.

Alternative valuation approaches are shown; both made use of the trading figures available.

In example 11.4A, the calculations assume that the value of the property is directly related to its ability to produce a trading profit. The capital value of the business as a going concern, including the freehold of the premises, is shown by this approach to be approximately £625 000. It will be seen that this is equal to a multiplier of the adjusted net profit of 5; looked at in terms of net receipts of £475 000, the multiplier is approximately 1:3.

Example 11.4B shows an alternative approach. The valuer expects to make a valuation by direct reference to rents and yields. By taking the trading information and proceeding by way of finding a divisible balance, which is then apportioned between the tenant and the landlord, there is a notional rental figure that can be used in a valuation as shown. The value by this method is within approximately 3 per cent of the previous valuation.

Following the above exercise, the valuer should look at the broader picture.

It is stated that the premises are situated in the retail centre of a market town. In forming a valuation, the valuer would need to look at a number of aspects. For example, how many other public houses operate in the centre? Are there too many for them all to be profitable? Is there scope for a full restaurant service; if so, could any first floor space be used to advantage? Is there a demand for retail premises and would one of the smaller multiples be interested in being represented in this town; alternatively, would one of the existing traders be anxious to move to gain a better trading position? What are the vacant possession values of retail shops in the area and how do these compare with the asking price?

Because of the good situation within the heart of the prosperous market town, it may be that the premises would be of interest to a national retailer. The rental value for retail may well be higher for the reason that the site location is more important to a retail business than to a public house and the situation of the building on a prominent corner site in the retail centre would create considerable interest if placed on the market. Furthermore, a letting to a national multiple offering a good covenant would ensure a lower yield and therefore a higher sale price.

Example 11.4C shows the effect of this scenario.

The important lesson from these variations is that the valuer should look at all options in advising the client.

12 The contractor's test

As has been seen, the valuation of a particular interest in land is normally made by reference to its tenure, its use and its income-producing capacity. When all factual information has been collected, the details of recent sales or lettings of similar types of property are analysed. The valuer is then able to make an informed judgement about the market value.

LACK OF MARKET EVIDENCE

However, not all valuations relate to market transactions: there are many instances of buildings or interests where an opinion about its value is required even though no market activity has taken or will take place. For example, a company may need a valuation of its property assets for inclusion in the company's final accounts, or an annual value may be required as the basis of a rating assessment. Such requirements do not normally present a problem because the valuer can draw on knowledge relating to sales and lettings of similar types of property.

A difficulty arises with certain types of buildings and uses where holdings rarely change hands. Examples include public and state schools, local authority holdings, chemical plants, oil refineries, gas works, sewage treatment works, airports, docks, power stations, holiday camps, sports and leisure centres, caravan sites, golf courses, cattle markets, shipyards, steelworks, and similar holdings. In short, because of the specialized nature of the building or the special attributes of the site, there is no likelihood that a capital or rental value could be assigned to the building on the basis of the analysis of comparable transactions.

As a consequence, there is no body of information enabling an investment type approach to be used. It will be noted from these examples that the holding does not necessarily have to be in public ownership or be one pursuing a monopoly activity, although these two categories are in the foreground of the need for a different approach.

Occasionally, a building of the type described above will be sold. However, the sale price will not assist in assigning a value to similar buildings in use, as the most likely reason for sale is that the building no longer fulfils the purpose for which it was built. It may have become redundant in the overall process or been superseded by the advance of technology or streamlining of activities. For example, a pumping station belonging to a water supply company may be unable to cope with current demands and be taken out of service and replaced by supplies pumped from a larger and more efficient source. The company, having no further operational use for the premises, will first consider whether they are of use for some other activity of the company, such as storage or workshop facilities. If no such use is identified, it will seek a purchaser at the best price available.

It may be that an alternative use is obvious and one for which planning permission would be readily forthcoming. If that is the case, the company would be advised to obtain outline planning permission and market the site with the benefit of the change of use. Where the future use is problematical, the site would probably be placed on the market seeking offers and leaving the prospective purchaser to consider the viability of any proposed future use in the context of the likelihood of obtaining planning permission and the degree of competition anticipated. Such uncertainties would tend to conspire to produce a very modest offer.

AN ALTERNATIVE APPROACH

Where the courts are engaged in cases where there is a valuation element, they have made it clear that it is not their function to look for a particular approach. It was said in *Garton v. Hunter (VO) (1969)* where the case concerned an assessment for rating: 'We admit all relevant evidence. The goodness or badness of it goes to weight and not to admissibility.' And also:

> We do not look upon any of these tests as being either a "right" method or a "wrong" method of valuation; all three are means to the same end; all [three] are legitimate ways of seeking to arrive at a rental figure that would correspond with an actual market rent on the statutory hypothesis and if they are applied all the tests should in fact point to the same answer; but the greater the margin for error in any particular test, the less is the weight that can be attached to it.

THE TEST

This chapter and the next are concerned with a consideration of the problems and a description of the contractor's test, a well-established solution to the valuation issues surrounding the types of property exemplified above.

Because of the absence of a clear base and the level of unsupported opinion thereby imported into such valuations, the test has been developed in a way that is quite different from other valuation methods. It is concerned with finding an annual or capital sum without the availability of market evidence.

To be credible, it is essential that the test produces an equitable result when used to assess a net annual value for rating revenue purposes or compensation payable where land is acquired under compulsory powers. The test may also be used to provide estimates of capital worth for balance sheet purposes.

The purpose of the test is to establish a value and this is achieved by estimating the construction cost of the building and the value of the land, making allowances for obsolescence and disabilities where appropriate.

The basic process is demonstrated in the following example designed to show the main elements of the test.

Example 12.1

An electricity supply company owns a high voltage sub-station erected some 60 years ago in an enclosed hill farm and adjoining moorland, the site extending to about 2 ha. It is approached over a lengthy unmade right of way. A valuation is required for internal financial purposes only; it is anticipated that the facility will continue in use for the foreseeable future.

estimated cost of construction	£4 500 000
value of land – current use	£ 20 000
total	£4 520 000
deduct for age and disabilities, 70%	£3 164 000
balance sheet value on contractor's basis	£1 356 000

Commentary

It is assumed that the equipment has been maintained to current standards. The installation is performing the task for which it was established.

However, the site was developed some 60 years ago, is in a remote location and has a sub-standard access (it is assumed that the company has no power to widen or otherwise improve the track without permission from the owner or owners). The combination of factors justifies a substantial deduction for age and obsolescence.

There is little likelihood that a vacated site could be put to an alternative economic use. The costs of integrating the site with the surrounding agricultural uses could not be justified financially.

The valuation shows a discount of 70%: even so, the figure is considerably greater than would be likely to be achieved should the site be decommissioned and offered for sale.

APPROACH TO ESTIMATING COSTS

The arbitrary nature of the basis, the use of cost or adjusted cost as a surrogate for value and the absence of any element of market value evidence all point to considerable scope for differences of opinion. The differing views of the parties may be held sincerely but need to be resolved in the process of which the build up of the figures will come under scrutiny.

A more detailed explanation of the costing process is given below.

Building costs

The building may be of a modern, functional form, in which case costings would normally be based on a replacement basis. Where the building was erected some time ago and in an elaborate style that would not be contemplated today, the appropriate basis would be on the assumption of the provision of a substituted building.

Replacement building

A replacement building would be one of similar appearance, footprint and height, although original architectural embellishment would be excluded. The cost would then be based on that of a broadly identical building using modern building techniques and materials insofar as they are compatible with the concept of replacement.

If the building is listed as being of historic or architectural interest, there would be little scope for change, the basis used would therefore be replacement. Any listed building is likely to inhibit the user in some way and the opportunity to effect any changes is extremely limited. The higher costs of any permitted work of alteration, improvement or maintenance add to the problem of finding an appropriate value based on the contractor's test. Many companies would find the restrictions limiting in their pursuit of business, although others would welcome the opportunity to occupy a building of distinction. In appropriate cases, the disadvantages judged to exist could be dealt with by an end-allowance for disabilities, as discussed below.

The estimated replacement cost would need to reflect not only the cost of reproducing the principal features of the building, but also the increased expense of maintenance. Additional research would be necessary to arrive at a reliable estimate of cost five main approaches are available..

Substituted building

The substituted building envisaged would normally be of modern design, taking account of any advances in technique since the original was erected. For example, a single-storey building may be regarded as a suitable replacement for a three-storey original building, with a row of stabling replaced by car-parking provision.

Cost estimates

In either case, the current cost will be estimated as accurately as possible, to include the building, external and site works, finance costs and fees.

The estimated cost of the replacement building will prove the more difficult to assess since elements of the building may not represent current practice. It should be possible to make a fairly accurate estimate of cost based on one of the following approaches or, in some combinations, elements taken from more than one of the approaches. Five main approaches to costing are available.

Superficial area

By far the most common method of estimating approximate building costs is the superficial area method. The building is measured in accordance with the relevant part of the Royal Institution of Chartered Surveyors' (RICS) *Code of Measuring Practice*. The resultant unit price (normally in cost per square metre) is applied to the area, with lump sum additions for stores and other structures outside the main building, landscaping and other site works, and car parking, to which are added building finance and professional fees.

The effect of plan shape on cost is not reflected directly in this approach, and the valuer should be aware that departure from a simple square or rectangular shape is likely to increase the cost.

Cubic content

Certain buildings may lend themselves more readily to a calculation of building cost using the total cubic content of the unit. The height will of course influence the final estimate. In this approach, the difference between costs for a flat roof and one with a pitched design are reflected by accepted but arbitrary allowances to the measured height of the building.

Elemental costs

In assessing costs by reference to the elements making up the building, the material and labour costs for each main component of the building are identified and totalled to arrive at an overall cost. The approach has gained in popularity since the adoption of cost-planning by leading practices of quantity surveyors. It requires a more detailed knowledge of construction than either of the two foregoing methods and its collation is likely to be more time-consuming.

Unit costs

A unit cost may be used for the approximate costing of, for example, schools, cinemas and hotels. Schools may be costed by comparing the price per pupil desk, cinemas by the cost per seat and hotels by the cost per bedroom. Such an approach is unlikely to be helpful except where the valuer is intimately involved with that particular type of building and has a wide knowledge of costs in that sector. The figure will, of course, have regional variations as well as differences relating to the quality of the building and its finishes.

Approximate quantities

Quantities may be taken off, priced and totalled. This is more easily carried out if detailed plans are available. There would be a significant cost to such an exercise that is unlikely to be within the experience or competence of a valuation surveyor.

There are several published sources of building cost information: Appendix A shows an edited extract from *Spon's Architects' and Builders' Price Book*.

The cost of construction is an approximation used in a method that is acknowledged as being indicative rather than precise. The exercise is carried out on a regular basis and despite the obstacles it is usually possible to arrive at a figure that is acceptable for the purpose.

ADJUSTMENTS TO BUILDING COSTS

Some adjustments may need to be made to reflect age, inconvenience and disabilities.

Depreciation

In arriving at a replacement value, some allowance may need to be made for depreciation. Depreciation may be defined as the measure of wearing out, consumption, or other reduction in the useful economic life of a fixed asset, whether arising from use, effluxion of time or obsolescence through technological or market changes.

It is common to divide the types of depreciation into more precise categories of obsolescence, as follows.

Physical or economic obsolescence

The wear and tear normally associated with age, which will extend to the additional periodic maintenance costs likely to be incurred.

Functional obsolescence

Functional obsolescence is defined as inadequacy of the design in terms of its current usage. Examples are unnecessarily high ceilings, redundant corridors and unsatisfactory layout.

Technical obsolescence

An extension of functional obsolescence where there are significant disabilities.

Market obsolescence

Market obsolescence relates to changes in operations rendering the layout or design redundant.

It is sometimes suggested that one or more of the methods of depreciation used by accountants in preparing business accounts should be adopted. Most of these measures are used to depreciate machinery or equipment used in the business and that have a relatively short life. The useful life can be estimated with a fair degree of precision and will help to determine the rate of depreciation. Such methods do not transfer easily to buildings, and in practice the allowance for obsolescence in the case of property is not likely to cause major problems.

It has been the custom to apply a depreciation factor to the combined total of estimated land and building costs. While this is generally adequate, it is suggested that the preferred approach is to deduct any allowance from the land and buildings separately. In some instances it may be essential. For example, a listed building would be likely to attract a high rate of depreciation, the special reasons for which would not apply to the land.

By showing separate depreciation figures for the land and buildings, there is an assurance that the two items have each received consideration.

Disabilities

Any disabilities not covered in the above categories should be reflected in an end deduction.

COST OF SITE

It is accepted that the cost of the land is fixed broadly with reference to the value of adjoining land. The reason is that it would not be possible to negotiate the purchase of a suitable site at a price less than could be achieved by selling it for whatever alternative use would be likely to be approved by the planners. In some cases, the alternative use would be obvious, while in others it could be unclear without some guidance. In example 12.1, the alternative use seems to be associated with rural land management. Without powers of compulsory purchase, the price would be subject to negotiation based on a figure somewhat above moorland value, with the vendor free to discontinue the negotiations, even where a good price was proposed. Where compulsory powers exist, there is less scope. But in either case, the owner could expect a price reflecting an increment over the current use value of the land, with possible additions where its loss would create special problems for the vendor or affect the value of the land retained.

Disabilities

As stated, the land would normally be valued on the basis of its alternative use, which would normally be apparent from the use of adjoining land.

In the case of land contaminated by the current use, the site may need some expenditure before it would be suitable for the alternative use of the adjoining land against which it would be valued.

THE OUTCOME

Various terms have been used to describe the final figure after the various adjustments and allowances – 'effective capital value' was used for a long time until the term was criticized by a tribunal member who favoured 'estimated replacement cost'. The current preference is for 'depreciated replacement cost' (DRC), which fairly describes the end result.

DRC is defined in the RICS guidance notes on this topic as:

- the open market value of the land for its existing use
- the current gross replacement cost of the buildings and their site works, less an allowance for all appropriate factors such as age, condition, and functional and environmental obsolescence, which all result in the existing property being worth less than a new replacement.

In asset valuations, the valuer must consult the directors of the company regarding trading.

In the event of disagreements of a fundamental nature that cannot be resolved, reference will be made initially, in the case of rating assessments, to local valuation courts. The Lands Tribunal will deal with appeals on rating assessments and with compulsory purchase cases. The decision of the tribunal is final on matters of fact but matters of law may be appealed to the Court of Appeal and then to the House of Lords.

The tribunal is a specialist court consisting of members with legal or technical (specifically valuation) qualifications and is made up of lawyers and surveyors. It hears and decides on technical evidence or interpretation of the application of the law. The technicalities in dispute are explored by one or more members depending on the complexity of the case and the nature of the dispute. The judgments are particularly helpful in providing guidance on the interpretation of the law and its effect on value. Its decisions cannot be appealed except on a point of law.

JUDICIAL GUIDANCE

The arbitrary nature of the basis, the use of cost or adjusted cost as a surrogate for value, the absence of any element of market value evidence and the reluctance of the judiciary to contemplate valuations offered on this basis make the process of arriving at a result more problematic. Where the test is used to value relevant properties for statutory purposes, lack of agreement between the parties has led to intervention over the years by the Lands Tribunal and the courts. The differing views of the parties may be held sincerely but need to be resolved, in the process of which the build up of the figures will come under scrutiny.

The judiciary has shown a general antipathy in considering cases where the figures in dispute have been arrived at using the contractor's test. But there is a general, albeit grudging, acknowledgement that there are some cases where there is no alternative to the test.

Both the Lands Tribunal and the courts have expressed considerable reservations about the test, which gives full rein to the valuer's judgement without offering any market-related evidence in support.

In justifying its resistance to the test, the Lands Tribunal has emphasized that in general valuation practice, the valuer identifies and isolates the characteristics of other transactions that have an effect on value and then uses that information in forming a view about the value of the property under consideration. These elements are not present in the contractor's test, placing greater emphasis on the standing and experience of the valuer and the need for integrity and independence in assembling the various components and reaching a conclusion.

In particular, the lack of any correlation between cost and value has provoked much criticism of the test. It has been described as 'a method of last resort', a view to which most valuers would raise no objection, preferring a more direct and referential approach where possible. Few practitioners would go beyond claiming that the test sets a negotiating platform rather than producing a precise value.

Despite being described as a method of last resort, the test is necessarily in use in the limited circumstances described and offers some form of ground rules for the valuation of a limited class of property. The main use for the test for annual values is in the hypothetical world of rating assessment; it is also used in assessing capital values in relevant cases of compulsory purchase, and for the assessment of certain company assets (the latter not being within the purview of the courts or Lands Tribunal).

The Lands Tribunal has also criticized the test for its 'artificiality of approach'. This misses the point in that the test is intended for use only in circumstances where the more usual valuation practices cannot be called upon. It has described it as 'a poor best' (which is no more than is claimed for it; the test is unlikely to be used where a viable alternative method is available). At the same time, it has been acknowledged that 'it is sometimes the only practical approach' (*Dawkins v. Royal Leamington Spa Corporation (1961)*).

In a rating case, the court observed: 'Where better evidence is in the circumstances of a particular hereditament impossible, resort may be had to either capital value or cost of construction' (*Robinson (Brewers) Ltd v. Boughton and Chester le Street (1937)*).

But the tribunal has also given some positive support to the test, suggesting in *Eton College v. Lane (VO) (1971)*: 'the time has come for a good word to be said for this method of valuation' and 'we are satisfied that the contractor's basis provides a valuation instrument at least as precise as any other approach'.

The tribunal member went on to emphasize the importance of the valuer's role, saying:

> Provided a valuer using this approach is sufficiently experienced and is aware of what he is doing and knows just how he is using his particular variant of

the method, and provided he constantly keeps in mind what he is comparing with what, we are satisfied that the contractor's basis provides a valuation instrument at least as precise as any other approach. For this kind of case it is almost certainly the best substantive method that has been devised so far.

In *Gilmore (VO) v. Baker-Carr (1963)* the Lands Tribunal enunciated a set of rules presented as five stages in the valuation process, which have found general favour:

- Estimate the cost of construction of the building.
- Distinguish between value and cost and make deductions from the cost of construction to allow for age, obsolescence and any other factors necessary to arrive at the effective capital value.
- Establish the cost of the land.
- Apply the market rates to decapitalize the capital value arrived at.
- Take account of any items not already considered.

A sixth stage was added by tribunal member Mr C. R. Mallett, who said:

> I hesitate to add another stage to the saga of the contractor's test but I think it is logical and necessary first to determine the annual equivalent of the likely capital cost to the hypothetical tenant on the assumption that he sought to provide his own premises, then to consider if the figure is likely to be pushed up or down in the negotiations between a hypothetical landlord and a hypothetical tenant having regard to the relative bargaining strengths of the parties.
> *Imperial College of Science and Technology v. Ebdon (VO) and Westminster City Council (1984)*

In *Cardiff City Council v. Williams (VO) (1973),* Lord Denning said:

> Even where the contractor's basis is taken, the assessment on that basis is open to great variations up and down, as, for instance, in assessing the effective capital value the possible variations may become so great that the contractor's basis ceases to be a significant factor ... In such a situation the tribunal may prefer to take some other basis ...

In the case of a large public school, it was conceded that the contractor's test was the best substantive method, subject to the proviso that the valuer using the method was sufficiently experienced, since it was the only evidence available (*Shrewsbury School (Governors) v. Hudd (VO) (1966)*). It is interesting to note that in this case the tribunal expressed a preference for the contractor's test over a valuation based on a value per student place.

Decapitalization

Historically, in the case of rating assessments, once the capital value had been assessed, the need for an annual value had to be considered. Where an annual value was required, the valuer would determine what appeared to be the decapitalization rate in the circumstances, recognizing that the rate would tend to differ according to whether the use could be described as commercial or not (the latter encompassing utilities, schools and the like).

The Lands Tribunal and the courts had developed what might be termed the decapitalization doctrine by defining the way in which a capital value arrived at by use of the test would be represented in annual terms. Although the legislation now makes provision for the rates at which capital sums are to be decapitalized, there is still some merit in following the underlying reasoning in approaching the question.

Since 1990 the percentage rates have been prescribed by statutory instrument. Initially, it was provided that the rate would be 4 per cent in the case of educational use or hospitals, and 6 per cent in all other cases. The latest provisions are 3.3 per cent for education, healthcare and defence establishments, and 5 per cent for all others (Non-Domestic Rating (Miscellaneous Provisions) (Amendment) (England) Regulations 2004 SI No. 1494).

It will be seen that there is a distinction between what could be described as social or public uses, and commercial uses. The preferential treatment of defence land is a later addition, suggesting that other property in public ownership could be included at some later date. It will be seen that a constant differential has been retained: in each case the 'concessionary' rate is two-thirds of the standard rate. The latter is fixed below the commercial rate of interest that would be regarded as appropriate.

The following account is a distillation of the evidence given and the conclusions reached in a number of cases where the further step of decapitalization was necessary in order to arrive at the annual value, almost always for rating purposes.

In a very early case, the court was satisfied that the interest on a loan to fund the purchase of a canal could be regarded as equivalent to rent (*R. v. Chaplin, 1831*).

However, in 1886, Cave J took a more analytical view of the equivalence of rent and interest:

> interest on cost is a rough test undoubtedly. It is a test in some cases but not a test in others. If the place is occupied by a tenant, it is not a good test at all, because the rent which he actually pays is a far better one … But if the place is occupied by the owner himself, then it is … in some cases a test, a rough test no doubt and only prima facie evidence, but still some evidence, to show what the value of the occupation is if he could get the place cheaper, at a less rent than the interest on the cost comes to, it is to be assumed he would not go to the expense of building, he would prefer to take the cheaper course and pay the rent.
>
> *R. v. School Board for London, 1886*

It will be noted that he had no problem with the use of the test.

In a more recent case (*Governors of Shrewsbury School v. Hudd (1966)*), Denning LJ (as he then was) was already expressing concern about the test:

Even when the contractor's basis is taken, the assessment on that basis is open to great variations up and down, as, for instance, in assessing the effective capital value and in deciding what percentage to take on it. The possible variations may become so great that the contractor's basis ceases to be a significant factor in the assessment. In such a situation the tribunal of fact may prefer to take some other basis.

Cardiff Rating Authority (1949)

He further asserted that, in the case of a school 'not run for profit', 'no tenant would be prepared to offer, nor would any landlord … expect … a rent based on a commercial percentage of the value of the school. The precise figure which would finally be reached is a matter of conjecture'.

This reinforced what was said in *R. v. Paddington (VO) ex parte Peachey Property Company Ltd (1965):* 'The rate of interest to be applied, as the criterion of rental value, is not what a contractor would ask but what he would get. It is the rate that would be arrived at after negotiation in the market.'

And also, quoting an earlier decision: 'The whole of the circumstances and conditions under which the owner has become the occupier must be taken into consideration, and no higher rent must be fixed as the basis of assessment than that which would finally be reached as a matter of conjecture.'

There have been attempts to build up the appropriate decapitalization rate by a theoretical approach, starting with the commercial market and making various adjustments. Three valuers gave evidence for the various parties involved in the Shrewsbury case: two arrived at 3.5 per cent and one at 6 per cent, which adds some weight to the criticism levelled against the test.

In another case, evidence was produced to show that in more than 100 agreed assessments of universities in England and Wales outside the old LCC areas, assessment was on the basis of a decapitalization rate of 3.5 per cent (all relating to a valuation date of 1 April 1973).

In *Dawkins v. Royal Leamington Spa Corporation (1961),* Sir Jocelyn Simon QC, appearing as Solicitor General, gave an elegant and authoritative description of the basis of fixing an annual value. He said, in part:

As I understand it, the argument is that the hypothetical tenant has an alternative to leasing the hereditament and paying rent for it; he can build a precisely similar building himself. He could borrow the money, on which he would have to pay interest; or use his own capital on which he would have to forgo interest, to put up a similar building for his own occupation rather than rent it, and he will do that rather than pay what he would regard as an excessive rent – that is, a rent which is greater than the interest he forgoes by using his own capital to build the building himself. The argument is that he will therefore be unwilling to pay more as an annual rent for a hereditament than it would cost him in the way of annual interest on the capital sum necessary to build a similar hereditament. On the other hand, if the annual rent demanded is fixed marginally below what it would cost him in the way of annual interest on the capital sum necessary to build a similar hereditament, it will be in his interest to rent the hereditament rather than build it.

In acknowledging this statement as the classic explanation, Lord Denning MR was also very specific in defining the relation of rent to interest, making the following qualification:

> The annual rent must not be fixed so as to be only 'marginally below' the interest charge. It must be fixed much below it and for this reason: by paying the interest charge on capital cost, he gets not only the use of the building for its life, but he gets the title to it, together with any appreciation in value due to inflation; whereas, by paying the annual rent, he only gets the use of the building from year to year – without any title to it whatsoever – and without any benefit from inflation.
>
> *Cardiff City Council v. Williams (VO) (1973)*

Mr C.R. Mallett, a tribunal member in the Imperial College case (above) indicated that the decapitalized rate was not the same as the market rate. He listed the following factors cited by Lord Denning in considering the likely level of rent to be paid, underlining the fact that various decisions are likely to come into play in the mind of the hypothetical tenant:

- The hypothetical landlord would be anxious not to leave the premises empty, bringing in no return on his capital. He would be content to accept much less than the 6 per cent that the landlord of commercial premises would expect.
- The hypothetical landlord would be providing the premises for the public good, i.e. for educational purposes, and would be disposed to accept such rent as the hypothetical tenant could afford to pay without crippling his or her activities, especially if it was a non-profit making body.
- On the one hand, the hypothetical tenant might be very anxious to get possession of the premises. If he was under a statutory compulsion to establish such a college, he would feel that he must have them – not perhaps at any cost, because if the rent was too high he would look somewhere else – but at such reasonable rent as he could persuade the hypothetical landlord to accept. If he felt he had a moral compulsion to establish such a college – out of his public duty – he would probably pay as much as if he were under a statutory duty.
- On the other hand, the hypothetical tenant might wish to have the premises – as being desirable – but not as being compulsory, not as a matter of urgency. He would be inclined, therefore, to hold back and pay less than a hypothetical tenant who felt a pressing need to get them.
- The hypothetical tenant would, of course, have to consider the means at his disposal. If he was a government establishment, such as a local authority,

supported entirely by government grants and the rates, he would count the cost, but perhaps not as anxiously as a non-government establishment under a pressing necessity to make both ends meet (*Cardiff City Council v. Williams (VO) (1973)*).

He (the tribunal member) went on:

I think it is logical and necessary first to determine the annual equivalent of the likely capital cost to the hypothetical tenant on the assumption that he sought to provide his own premises, then to consider if the figure is likely to be pushed up or down in the negotiations between a hypothetical landlord and a hypothetical tenant having regard to the relative bargaining strengths of the parties.

In a typically down-to-earth comment, Lord Denning MR suggests that attempting too much precision is unrealistic. He said: 'whether it be 3.5 per cent or 4.5 per cent is all guess work anyway – you are working in a void'.

As stated above, the allowances are now embedded in legislation, providing an equality of treatment over the whole of England and Wales.

The examples in the following chapter shed more light on the approach taken by the tribunal and the courts in particular cases where the contractor's test has been deemed appropriate. To a large extent, they represent good practice, which a valuer would be well advised to take into account.

SUMMARY

- It is well recognized that cost and value bear little relationship.
- There are some instances where valuations are required of properties for which there is no market. In such cases, the cost of construction is estimated and reduced as necessary to reflect obsolescence.
- The land value is estimated on the basis of the use of adjoining land.
- Care and experience are necessary to ensure that a sensible result is achieved.
- Valuations on this basis may be used for compulsory purchase and rating requirements: also for asset values and other mainly internal accounting purposes.

13 Application of the contractor's test

INTRODUCTION

It is now possible to consider the practical application of the principles described in the previous chapter. A brief account of the circumstances of each example is set out, followed by a summary of the way in which the valuation was built up. The subsequent commentary draws attention to significant aspects of the decision or result.

The first example demonstrates why a commercial rate of interest would not be appropriate in finding an annual value from the cost information assembled for the exercise. Part A shows the value of existing premises, while Part B applies the process to newly built premises.

Examples 13.2 to 13.7 are based on actual decisions of the Lands Tribunal or the courts and disclose a wide range of approaches: in general, there appears to be a preference for a common sense rather than an academic approach. But the starting point is the accuracy and integrity of the information used so far as this is possible.

Most of the examples quoted relate to valuations for rating. Specialist valuations are dependent not only on the particular method employed, but also on the constraints under which they are applied. This text is restricted to a description of the method and does not attempt to describe the applications in any detail as they are subject to complex statutory provisions and beyond the scope of the present book. Further help may be obtained from some of the texts listed in Further reading.

EXAMPLES

Example 13.1 A regional water board

A major regional water board has recently completed the building of operational premises for its own use. The cost of the land was £500 000 and that of the building £3 000 000. The development is of the same size and fulfils the same purpose as an adjoining building erected 60 years ago. The new building has been provided to cope with additional demand. The original building will continue in full operation.

Using the contractor's test, an experienced valuer has arrived at a figure of £1 750 000 for the existing building after making a 50 per cent reduction for age and obsolescence generally. The board has a policy of owning all its buildings and land. Major capital works are financed from an internal sinking fund set up for the purpose.

The following example demonstrates the separate approaches to the two sets of premises.

Valuation A - Existing premises

			£
Cost of erection of building, site works, etc.			3 000 000
Add	Land value for current use		500 000
		Total	3 500 000
Less	Obsolescence at 50%		1 750 000
			1 750 000
	Annual value at 5%	87 500	

Valuation B - New premises

			£
Cost of erection of building, site works, etc.			3 000 000
Add	Land value for current use		500 000
		Total	3 500 000
	Annual value at 5%	175 000	

Commentary

It is worthwhile using this example to consider whether it would be possible to decapitalize by using or adjusting market rates, as suggested by some observers.

If we put aside for the moment the reason for using the contractor's test to find a 'market' figure, which is that there is no market, let us assume that the company wishes to hold its operational premises on lease to release capital for use in the business and that there is an investor willing to purchase this type of property and to let it to the operator, knowing that there is no other market for the premises. The motivation of any investor will be to ensure a return on the investment commensurate with the risk. We assume that the investor would wish to enter into separate commercially-based leasebacks and would look for a higher yield on the older property to reflect the realities of the situation.

If it is determined that the modern building would command a yield of 8 per cent, a suitable yield for the older building would be in the order of 12 per cent. On the figures assigned above, this would give rents or returns of £280 000 and £210 000 respectively. This immediately highlights an anomaly; the capital sums found using the contractor's test are in the proportion of 2:1, whereas the rents based on market yields show a ratio of 4:3 and therefore erode the benefit to the company. This places any business with capital reserves in a favoured position. This is, of course, attributable to the compensation provided by the yield differential, but can hardly be equitable.

The decision to take away part of the valuation process by fixing the decapitalization rates by legislation removes the possibility of the anomaly described above. The obsolescence recognized in the capital amounts is fully reflected in the converted rental figures.

Example 13.2 Independent day school

(Based on *Westminster City Council v. American School in London and Goodwin (VO) (1980)*)

Both parties agreed that the school (established by the American community after the Second World War to offer a system of education geared to that pertaining in the United States of America) was to be valued on the contractor's basis.

The effective capital value was agreed at £2 350 000 'to reflect any age or obsolescence allowance that would fall to be considered in the valuation on this basis but not any end allowance'.

The only matters at issue were therefore the appropriate percentage rate to be applied to the capital value to produce an annual value, and the amount of any end allowance.

The tribunal decided to assume an interest rate of 5 per cent (rather higher than the customary rate of 3.5 per cent widely applied to this class of building at that time). It was noted that the agreed effective capital value did not include any end allowance for disabilities; a deduction of 6 per cent was made for this item.

		£
Capital value reflecting obsolescence (agreed by the parties)		2 350 000
Convert to annual terms at 5%	117 500	
Deduct End allowance for further disabilities – 6%	7 050	
	110 450	
Gross value, say, 110 000		

The same result would have been obtained by deducting the end allowance from the capital sum and then converting to annual terms as shown here. It has the advantage of being easier to follow the reasoning of the tribunal.

Capital value reflecting obsolescence (agreed by the parties]	2 350 000
Deduct To allow for disabilities – 6%	141 000
	2 209 000
Convert to annual terms @ 5%	110 450
Gross value, say (as before), 110 000	

Commentary

Although the case was a celebrated one and the report detailed, the valuation calculation was of the simplest form. The capital value adduced, in the view of the tribunal, already reflected some of the disabilities. The tribunal noted a range of disabilities, including cramped site, high density with limited parking, shortage of playing fields, swimming pools and other recreational facilities, seven different floor levels on the ground floor and stairs and ramps due to the presence of both a sewer and a railway beneath the site. While, in its view, some of the disabilities were already taken into account in arriving at the effective capital value, it made a further deduction of 6 per cent as an end allowance, equivalent to a deduction of £141 000 from the capital value.

The figures used in the decision have been recast to show the effect of these comments.

Example 13.3 A members' cricket club

(Based on *Marylebone Cricket Club v. Morley (VO) (1959)*)

The Lands Tribunal considered a dispute arising about the proper valuation for rating of the world-famous Lord's cricket ground.

The ratepayers' valuer initially approached the valuation on the profits basis, but detailed figures showed that a loss was made; he contended for an assessment to rateable value of £3500 (which by coincidence was the approximate rateable value from 1935–56).

The valuation officer presented two valuations on the contractor's basis, the first made by a contract valuer.

The two valuations had been made by different members of the valuation office

Valuation A

				£
Land Playing area	5.25 acres at £4 000	21 000	at 4%	840
Nursery field	5.75 acres at £2 000	11 500	at 4%	460
Car parks	2.5 acres at £275	687	at 4%	34
				1 334
Add	Buildings	124 247	at 5%	6 212
				7 546

Say £7 500 annual value

Valuation B

Capital value for the purposes for which the land is used		
16.75 acres at £6 000 per acre	100 500	
Annual equivalent at 4%		4 020
Add Building and site works: cost after making		
allowance for excess space, height, etc.	409 380	
Annual equivalent at 5%		20 469
Total annual equivalent of land and buildings		24 489
Deduct One-third for disabilities		8 163
Adjusted annual equivalent		16 326

The difference between the two valuations was remarkable, said the tribunal, because although they were 'of the same property, on the same date, for the same purposes and on the same side' they were so different in amount.

The tribunal (Mr John Watson) fixed the rateable value of the hereditament at £9000, but in doing so rejected the use of the contractor's basis on the following grounds:

> we accept [the] proposition that where a hereditament has only one possible tenant its rental value is restricted by the ability of that tenant to pay. That is where the contractor's basis breaks down … where the tenant is pleading inability to pay in rent anything approaching what the contractor's method would appear to indicate, the proposition postulates an enquiry into how he is conducting his affairs within the economic limits his policy has imposed. Without abrogating that policy, has he used every endeavour to maintain and increase his income on the one hand and curtail his expenditure on the other? For the purpose of such an enquiry it is well established in law that events the tenant could reasonably have foreseen at the date of the valuation, and the effect they are likely to have had upon his mind, are admissible in evidence.

Commentary

The figures in Valuation A presumably incorporate obsolescence, although there is no specific reference thereto. Further, it identified only net areas of land assigned to specific purposes (playing fields, nursery field, car parks), ignoring land occupied by buildings or used for circulation. As a result, only 80 per cent of the total site area was accounted for.

Valuation B makes an allowance of one-third of the total annual equivalent for obsolescence. The deductions are made after the rental values have been derived from the capital costs assigned to the land and buildings. A further allowance of one-third is made for disabilities, but still produces an annual value more than twice that of the first valuation.

The valuation evidence is unsatisfactory in several respects. There was no explanation of the wide variation between the two valuations produced.

The cricket club is a private members' club, its income depending on investments, attendance receipts and members' subscriptions. It is not operated primarily as a commercial organization and no enquiry seems to have been made of the way in which the annual budget was determined. As a private members' club, there was no imperative to make a profit, suggesting a need for a more searching investigation of the basis of the club than appears to have taken place. There was no evidence that the club was in danger of closing, yet no one seems to have asked how a club with an annual loss nevertheless managed to continue in existence. Before the case reached the tribunal, it had been accepted that the profits approach was not a satisfactory basis.

The tribunal member accepted the tenant's argument that it was unable to afford an assessment based on the calculated rental value and came to the conclusion that the contractor's test should be rejected. The profits test had been rejected earlier by the parties. The outcome is unsatisfactory, there being no information about the way in which the tribunal arrived at the rateable value.

The case perhaps emphasizes the necessity of presenting clear and accurate information before a tribunal.

Example 13.4 A small regional airport

(Based on *Coppin (VO) v. East Midlands Airport Joint Committee (1971)*)

The airport is conveniently near the M1. At the time of the appeal the airport had new, purpose-built terminal buildings and a single main concrete runway. The expenditure on the latter had been greater than was necessary for handling the volume of air traffic at the time and was undertaken with future expansion in mind. The airport was run by a joint committee of local authorities in the region and they accepted that there would be an annual deficit on the cost of running the airport. For this reason, the valuation was not made using the profits basis: the tribunal member stated: 'there was no profit motive and no profit and therefore the profits basis was clearly inappropriate'.

Both valuers pursued the contractor's basis but produced widely differing results. (valuation officer, £37 000; valuer for respondent ratepayers, £11 400). The main differences were the decapitalization rate (5 per cent as opposed to 3.25 per cent) and the deduction for disabilities (15 per cent and 40 per cent, respectively).

The decision of the tribunal member (Mr J .H. Emlyn Jones) reconciled the views of the two valuers. He also accepted that some recognition should be given to what the valuation officer first termed 'under user' but later amended to 'new venture allowance', which he then incorporated into the allowance for disabilities.

While observing that he would normally accept 5 per cent as the proper rate of return, he preferred to acknowledge the 'new venture' aspect by adjusting the rate of interest to 3.75 per cent rather than include it as an end allowance under the heading of disabilities.

		£
Total capital cost		1 239 000
Less Cost of excess runway	103 070	
Proportion of cost of let portion	145 000	248 070
Relevant capital cost		990 930
Less Sum to reflect date of valuation 12.5%		123 866
		867 064
Less Sum for disabilities (including new venture allowance), say		67 064
		800 000
Effective annual value at 3.75%		30 000

Commentary

The case rehearsed the arguments for and against the use of the profits method and concluded that it was inappropriate in this instance. In adopting the contractor's method, there were allowances in respect of excess capacity and the 'new venture' status of the undertaking (the latter resulting in a 25 per cent reduction in the effective annual value).

The acceptance that there was no profit motive suggests that the member was looking at an actual occupier as opposed to the hypothetical occupier required by rating law. Some difficulty seems to lie in the possibility of a similar regional airport operated by a commercial company for profit – would the assessment be dealt with differently?

The valuation on which the annual value was based appears to have been generous towards the ratepayer. It may have been reasonable on the part of the tribunal member to acknowledge the innate problems of a new venture, but he then went on to reduce what he referred to as the proper decapitalization rate of 5 per cent by 25 per cent, which suggests some element of double counting. The current system prescribes the decapitalization rates, ensuring a uniform and fairer approach in such cases.

Example 13.5 A public school

(Based on *Governors of Shrewsbury School v. Hudd (VO) (1966)*)

In an appeal against the gross value placed on the school, the ratepayers' valuer built up the valuation by taking a unit price per 'equivalent boarder' (2.5 day boys being equivalent to one boarder); extra amounts were then added to the previous total for amenities not likely to be present in all public schools.

The valuation officer preferred to proceed by formulating a valuation using the contractor's basis. The valuer found an effective capital value as follows.

		£
Cost of modern substituted buildings		1 205 955
Less Average deduction of 60.5%		730 526
		475 429
Add Site works	47 543	
Professional fees	41 838	
Land	48 980	138 361
Effective capital value		613 790

The tribunal increased the allowance for obsolescence to 70% and determined a gross value of £15,750, based on a decapitalization rate of 3.5%.

The tribunal (Mr Erskine Sims and Mr H. P. Hobbs) felt that the valuation officer's starting figure was too high, and that in arriving at his deduction he relied too much upon 'guidelines produced and too little upon his own judgement'.

The decision was based on increasing the allowance to 70 per cent, taking 3.5 per cent to show a gross value of £15 750 for the school buildings included in the assessment.

The tribunal went on to make other adjustments that are not relevant to this account.

Commentary

The tribunal did not accept the unit price per boarder approach, presumably because there was no adequate comparison on which to form a view about the value of a place.

The buildings were old and unsuited and the cost of building was therefore estimated on the basis of modern substituted buildings. The deduction for obsolescence varied with the age and the state of individual buildings, but overall averaged just over 60 per cent. It should be noted that site works, professional fees and land were calculated separately and not subjected to the allowance made on the buildings.

The tribunal observed that the valuation officer's starting figure was too high. Surprisingly, given this statement, the members did not reduce the starting figure but increased the allowance for obsolescence to 70 per cent.

Example 13.6 A school within the University of London

(Based on *Imperial College of Science and Technology v. Ebdon (VO) and Westminster City Council (1984)*)

The valuation officer and the valuers for the other two parties all gave evidence. There was agreement that the contractor's basis was the most appropriate approach in ascertaining the assessment. There was only one difference of opinion about the estimated replacement cost of the 15 buildings involved, disagreement centring on the amount of the disability allowance. The parties ranged in their opinions from 0–54 per cent.

The member of the Lands Tribunal (Mr C. R. Mallett) summarized the evidence given about the appropriate allowances and then proceeded to use his discretion based on the evidence to determine the gross value for rating purposes.

The tribunal discussed its decision through the five stages suggested in *Gilmore (VO) v. Baker-Carr (1963)* and added a sixth stage aimed at enabling the figures produced by the process to be reviewed at the end of the exercise. The member said:

> I think it is necessary and logical first to determine the annual equivalent of the likely capital cost to the hypothetical tenant on the assumption that he sought to provide his own premises, then to consider if this figure is likely to be pushed up or down in the negotiations between a hypothetical landlord and a hypothetical tenant having regard to the relative bargaining strengths of the parties.

In the event the final figure of £767 205 was rounded to £767 000.

The Lands Tribunal decision was upheld by the Court of Appeal; leave to appeal to the House of Lords was refused.

The decision of the tribunal:

		£
Estimated replacement cost (agreed by the parties)		20 340 438
Add Capital value of land, 15.725 acres @ £400 000		6 290 000
		26 630 438
Deduct For obsolescence, 11%		2 929 348
		23 701 090
Deduct End allowance for disabilities, 7.5%		1 777 582
		21 923 508
Convert to gross value @ 3.5%	767 205	
Say	767 000	

(Figures rearranged to simplify the presentation.)

Commentary

Although the parties agreed that the contractor's basis was the correct approach to the valuation of the buildings, there was a wide divergence of views about the amount of the disability allowance, if any.

Example 13.7 Purpose-built high security cash centre

(Based on *Barclays Bank plc v. Gerdes (VO) (1987)*)
The building was of a warehouse-type erected in 1982, but it differed from normal warehouses in that it was fortified to provide a secure building as a collecting, sorting, storage and distribution point for notes and coins.

Although the assessments of 15 similar properties throughout England were presented in evidence by the ratepayers, the valuation officer found them of no assistance as he could detect no consistent pattern of values. He therefore valued the premises by reference to the cost of construction using the contractor's basis of valuation.

The following valuation was presented to the Lands Tribunal by the valuation officer.

Valuation presented by valuation officer:

		£
Actual cost of construction in 1982		788 854
Adjust for cost levels in 1973 (say 25%)	197 214	
Add professional fees at 12%	23 666	
Gross value of buildings and site works (1973)	220 879	
Deduct 20% for superfluity on specification	44 176	
	176 703	
Add Value of land for current use (in 1973)	25 000	
	201 703	
Gross value at 6%	12 102	say 12 000

Two points will assist in understanding the figures. First, because the figures were based on a known cost of construction in 1982, it was necessary to reduce the figure to reflect the 'tone of the valuation list' as at 1 April 1973. Second, there is a reference to 'superfluity' but no allowance. Originally the valuation officer had made a deduction for this item of 20 per cent of the effective capital value of the building in 1973, but he later decided that the deduction was not warranted. The allowance related to the need to install air-conditioning throughout the building because of the lack of windows. The failure to agree was rooted in the valuation officer's view that the building cost was approximately three times that of a standard warehouse unit, while the valuer for the ratepayers was contending for a gross value assessment only marginally in excess of the assessment of a standard unit.

The member of the tribunal rejected the use of the contractor's basis in this case and criticized it on two counts.

He said:

> I do not think that the conversion of an actual 1982–3 cost to an estimated 1973 cost can be made with any precision by reference to the change in the estimates of approximate general building costs over the years as extracted from various publications on building cost estimates. This must be particularly so when dealing with a type of building that did not exist in 1973.

He added:

> Second, in trying to arrive at the appropriate percentage in order to reduce capital value to annual value, I do not think that in the present case any proper consideration was given to the essential difference between capital cost and annual value in the case of adjoining standard warehouse units built by a developer on borrowed capital for speculative sale or letting, and in the case of the appeal premises, which were built for a bank for their own occupation, without any profit motive and with, presumably, an ability to make capital available on more favourable terms.

Commentary

The decision demonstrates the difficulty of assessing a modern building for rating where the use was for specialized high security purposes, although the unit gave the general appearance of a warehouse.

Example 13.8 A private hospital service in a converted house

You act for the directors of a private hospital service devoted to the treatment of patients suffering from terminal illness.

They occupy a large Georgian house in extensive grounds on the outskirts of a large city. The original floor area has been doubled by the addition of a series of single-storey extensions over a period of years.

You have received instructions to prepare an asset valuation for incorporation into the annual accounts.

		£
Cost of modern substitute building with same gross internal area		1 640 000
Add Fees, finance charges		474 000
		2 114 000
Deduct Obsolescence, –45%		951 300
		1 162 700
Add Open market value of land restricted to exisiting use (inclusive of costs)		250 000
Depreciated replacement cost		1 412 700

(Subject to the adequate potential profitability of the business to be certified by the directors.)

Commentary

The building is of a type that would be difficult to value by reference to the open market. It comprises an old converted building with modern additions and appears to have more land than necessary for its present purpose. There may be an opportunity to obtain planning permission to develop the whole or part of the property.

However, for the present purpose, a valuation of the premises to include in the annual accounts, these issues are irrelevant. The premises are to continue to be used as a private hospital.

The valuer has assumed a modern substitute building for which an estimated cost can be readily obtained and which is then discounted, heavily in this case, to reflect the relative unsuitability and inconvenience of the present set of buildings. A land value is then added, to arrive at the depreciated replacement cost. The value assigned to the land is limited to its value for the existing use.

When the final figure is presented to the directors, it is their responsibility to consider it and adjust it downwards if, in their opinion, the adequate profitability of the business is insufficient to support the use of the land and building assets for their present purpose.

Appendix A: Edited extract from *Spon's Architects' and Builders' Price Book*

A sample of costs per square metre for a range of buildings for different uses

Administrative buildings		From (£)	To (£)
County courts		1 550	1 930
Magistrates courts		1 180	1 500
Civic offices			
Non-air-conditioned		1 180	1 500
Fully air-conditioned		1 490	1 750
Probation/register offices		860	1 230
Offices			
Prestige, high-rise, air-conditioned, iconic speculative towers		2 000	2 500
For owner-occupation			
low-rise, air-conditioned		1 300	1 600
medium-rise, air-conditioned		1 770	2 080
high-rise, air-conditioned		2 000	2 500
For letting			
Low-rise, air-conditioned, high quality speculative		1 200	1 510
Medium-rise, air-conditioned, high quality speculative, 8–10 storeys		1 460	2 000
Medium-rise, air-conditioned, city fringe, deep plan speculative towers		1 770	2 000
On business park			
Functional, non-air-conditioned	Less than 2000 m²	730	940
Functional, non-air-conditioned	More than 2000 m²	630	890
Medium quality, non-air-conditioned	Less than 2000 m²	830	1 030
Medium quality, non-air-conditioned	More than 2000 m²	780	1 000
Medium quality, air-conditioned	Less than 2000 m²	940	1 100
Medium quality, air-conditioned	More than 2000 m²	890	1 100
Good quality, naturally ventilated	Less than 2000 m²	990	1 100
Good quality, to BCO specification	More than 2000 m²	940	1 100
High quality, air-conditioned	Less than 2000 m²	990	1 300
High quality, air-conditioned	More than 2000 m²	940	1 350

Commercial premises		From (£)	To (£)
Banks			
Local		1 380	1 720
City centre/head office		1 970	2 540
Building societies			
Branch offices		1 250	1 630
Refurbishment		700	1 220
Shops			
Shells			
Small		590	760
Large, including department stores and supermarkets		510	720
Industrial-type buildings			
Shell with heating to office areas only	500–1000 m²	310	730
	1000–2000 m²	240	650
Unit including services to production area	500–1000 m²	560	870
	1000–2000 m²	500	800
	Greater than 2000 m²	500	800
Light industrial/offices			
Econmical shell and core with heating only		510	890
Medium shell and core with heating and ventilation		790	1 170
High quality shell and core with air-conditioning		1 040	1 900
Developer's Category A fit-out		420	720
Tenant's Category B fit-out		190	560
Factories			
For letting, shell and core only		560	760
For owner-occupation, controlled-environment, fully finished		1 200	1 600
Factory/office buildings (high technology production)			
For letting, shell and core only		560	760
For owner-occupation, controlled-environment, fully finished		1 200	1 600
Warehouse and distribution centres			
High bay (10–15 m) for owner-occupation (no heating)	Up to 10 000 m²	280	380
High bay (10–15 m) for owner-occupation (no heating)	10 000–20 000 m²		280
High bay (16–24 m) for owner-occupation (no heating)	10 000–20 000 m²		400
High bay (16–24 m) for owner-occupation (no heating)	Over 20 000 m²		360
Laboratory workshops and offices		1 080	1 360
High-technology laboratory workshop, air-conditioned		2 520	3 230
Agricultural storage buildings		450	760

Further reading

Baum, A. and Crosby, N. (1998) *Property Investment and Appraisal*, Routledge, London.

Bowcock, P. (1978) *Property Valuation Tables*, Macmillan Press, London.

Byrne, P. (1996) *Risk, Uncertainty and Decision-making in Property Development*, E&FN Spon, London.

Davidson, A.W. (2002) *Parry's Valuation and Investment Tables*, Estates Gazette, London.

Donaldsons (1988) *Donaldsons Investment Tables*, Donaldsons, London.

Freedman, P., Shapiro, E. and Steele, K. (2006) *Business Lease Renewals: the New Law and Practice*, Estates Gazette, London.

Laing and Buisson, *Guide to the Healthcare Industry*.

Langdon, D. (2008) *Spon's Architects' and Builders' Price Book*, E&FN Spon, London.

Hayward, R. (ed.) (2008) *Valuation: Principles into Practice*, Estates Gazette, London.

RICS (2007) *Code of Measuring Practice: a guide for property professionals*, RICS, London.

RICS (2008) *RICS Appraisal and Valuation Standards (The Red Book)*, RICS, London.

Index of statutes and statutory instruments

Index of cases

Subject index